Agriculture in the Forest-Steppe Region of Khazaria

East Central and Eastern Europe in the Middle Ages, 450–1450

General Editors

Florin Curta and Dušan Zupka

VOLUME 66

The titles published in this series are listed at *brill.com/ecee*

Agriculture in the Forest-Steppe Region of Khazaria

By

Volodymyr Koloda
Serhiy Gorbanenko

Translated by

Maryna Sergeyeva

BRILL

LEIDEN | BOSTON

Cover illustration: Mokhnach hillfort (Photo by Dmytro Yushkov, 2017).

Library of Congress Cataloging-in-Publication Data

Names: Koloda, Volodymyr, author. | Gorbanenko, Serhiy, author.
Title: Agriculture in the forest-steppe region of Khazaria / by Volodymyr Koloda, Serhiy Gorbanenko ; translated by Maryna Sergeyeva.
Description: Leiden ; Boston : Brill, 2020. | Series: East central and eastern Europe in the middle ages, 450-1450, 1872-8103 ; volume 66 | Includes bibliographical references and index.
Identifiers: LCCN 2020026206 (print) | LCCN 2020026207 (ebook) | ISBN 9789004349445 (paperback) | ISBN 9789004429574 (ebook)
Subjects: LCSH: Khazars—Agriculture—History—Middle ages, 600-1500. | Agriculture—History—Europe, Eastern—Middle ages, 600-1500.
Classification: LCC S469.E852 K65 2020 (print) | LCC S469.E852 (ebook) | DDC 630.939/8—dc23
LC record available at https://lccn.loc.gov/2020026206
LC ebook record available at https://lccn.loc.gov/2020026207

Typeface for the Latin, Greek, and Cyrillic scripts: "Brill". See and download: brill.com/brill-typeface.

ISSN 1872-8103
ISBN 978-90-04-34944-5 (paperback)
ISBN 978-90-04-42957-4 (e-book)

This book is printed on acid-free paper and produced in a sustainable manner.

Contents

Acknowledgements

This research has become possible due to many reasons. In the first place, the human factor was the most important. Therefore, the authors are grateful to all the specialists who collaborated with them. We are talking about various areas of research and methods in the natural sciences. First of all, the authors are grateful to Yuri Chendev (Belgorod State National Research University, Russia) for his long cooperation in the study of soils of antiquity, which lasted more than a decade and a half. We are also grateful to paleogeographers Zhanna Matviishina and Serhiy Karmazinenko (Institute of Geography, NAS of Ukraine, Kyiv, Ukraine) for the work with soils from the P'iatnyts'ke 1 settlement.

This monograph would not have been possible without the work of Roman Kroitor (Maison Méditerranéenne des Sciences de l'Homme, Aix-en-Provence, France), who significantly expanded zooarchaeological research with new data. Halyna Pashkevich (at that time Institute of Archeology of NAS of Ukraine) initiated other important definitions of palaeobotany. Thanks to Maryna Sergeyeva (Institute of Archeology of NAS of Ukraine) anthracological research on settlements of Saltiv culture was practically started.

This research would be impossible without work in various scientific organizations and institutions in Ukraine and Russia. Using this opportunity, the authors thank the staff of the Science Funds of IA, NAS of Ukraine, Science Archive of IA, NAS of Ukraine, Museum of Archeology and Ethnography of Slobidska Ukraine (Kharkiv, Ukraine), Art-memorial Museum of Ilya Repin (Chuhuiv, Ukraine), Archeological Museum of Voronezh State University (Russia), Divnogorie Natural Museum-Reserve of Architecture and Archaeology and many others.

Our opponents deserve special thanks. Konstantin Krasil'nikov became a reviewer of the first monograph of 2010 (see *Arkheolohiia*, 2011, 3). Yuri Pugolovok became a critic of a 2013 study (see *Arkheolohiia*, 2014, 3).

Volodymyr Koloda and Serhiy Gorbanenko

Illustrations

Diagrams

Figures

Maps

Photos

Plates

Introduction

An account of the early medieval history of the peoples of Eastern Europe is unconceivable without the Khazar Khaganate. During the period under study in this book, that was one of the largest polities in Eurasia. Through the integration of ethnically different peoples from the northern slopes of the Caucasus Mountains to the south to the Donets and Oskol rivers to the north, and from the Lower Volga to the east to the Dnieper and the Crimea in the west, the Khazar authorities succeeded in creating an economically powerful state, and managed to harness also the experience of craftsmen in the eastern Byzantine provinces. Because of that, towards the end of the first millennium, the Khazar Khaganate became one of the most powerful military and political forces in the Black Sea region, along with Byzantium, Kievan Rus', and the Arab Caliphate.

The history of research on the rise, development and decline of the Khazars, goes back several centuries. It has been described in detail in Peter Golden's studies.[1] In Russian historiography, the best survey remains that of Eduard Vashchenko.[2] The history of archaeological research has been presented by Svetlana Pletneva.[3]

Beginning with the 1990, and for the subsequent quarter of a century, the Kharkiv-based historian Oleksandr Tortika (1967–2015) dealt extensively with Khazar issues.[4] His studies were primarily based on written sources, with only occasional use of the archaeological data. It is important to note that in post-Soviet historiography, he had a unique position, as the only historian to focus on the Khazar Khaganate with so much purpose, and with such a broad thematic range. He placed a great deal of emphasis on military and political history, on social relations, as well as on ethnic and religious processes in Khazaria, all of which are reflected in his book.[5] In terms of economic history, Tortika dealt especially with trade routes passing through Khazaria and linking the European and Muslim markets.[6]

The most significant contributions to the study of Khazar history are those of such scholars as Douglas M. Dunlop, Omeljan Pritsak, Peter Golden, Norman Gold, R. A. E. Mason, and Constantin Zuckerman.[7] They all relied on written and numismatic sources. By contrast, in addition to written sources (known to him, at that time, in Russian translation), Mikhail I. Artamonov employed a considerable amount of archaeological data.[8] As all sites attributed to the Khazar period were within the USSR, foreign scholars had little knowledge of, and access to archaeological sources.

Khazaria occupied a large territory stretching over different climate and landscape zones, between the forest-steppe belt along the Don and the Dnieper rivers to the steppes along the Volga to the Lower Dnieper, to the Crimea, and, through the regions of Stavropol and Kuban, all the way to the northern foothills of the Caucasus Mountains. Such a diversity had a considerable impact upon the nature of economic activity and the mode of life of the constituent populations — sedentary in the forest-steppe zone and in the Caucasus, but nomadic or semi-nomadic in the steppe zone.

The ethnic diversity and various forms of reproductive economy, each one determined by a specific place of habitation, are responsible for the creation of a highly developed material culture, with considerable variations in its manifestations. Archaeologists have coined this the Saltiv (or Saltovo-Mayaki) culture, and have dated it between the mid-8th and the mid-10th century.[9] Svetlana Pletneva designated Saltiv (Saltovo-Mayaki) as the "state culture" of the Khazar Khaganate.[10]

Saltiv assemblages became the object of scholarly interest in the late 19th and the early 20th century, following the discovery of the Mayaki hillfort in 1890 and the excavation of the first catacomb graves in Verkhnyi Saltiv in 1900.[11] The scholarly interest in those assemblages increased slowly by steadily to the present. The reason for that favorable attitude to the history and material culture of the Khazars lies in their significant role not only in Eastern Europe, but also in Byzantium, the Black Sea, and Transcaucasia.

---

1 Golden 2005.

2 Vashchenko 2006.

3 Pletneva 1999.

4 Aksenov, Evseenko, and Riapolov 2016.

5 Tortika 2006b.

6 Tortika 2006a; 2006b; 2006c; Lobanova-Gulak and Tortika 2009.

7 Dunlop 1954; Pritsak 1981 and 1995; Golden 1983; Gold and Pritsak 1982; Mason 1995; Zuckerman 1995.

8 Artamonov 1962. Between 1934 and 1936, Artamonov headed the excavations in Sarkel and, after the war, those on the Right-Bank stronghold site at Tsimliansk (1949–1951).

9 The culture was named after Verkhnyi Saltiv (in the region of Kharkiv, Ukraine), the site where the first excavation of a cemetery with catacomb graves took place; and after Mayaki (in the region of Voronezh, Russia), the village next to which the first stronghold was investigated.

10 Pletneva 1999, 206–211.

11 Pokrovskii 1905; Makarenko 1906, 122–44; 1911; Miliutin 1909; Spitsyn 1909.

 | DOI:10.1163/9789004429574_002

Not all peoples included into the Khaganate, who participated in the creation of the Saltiv archaeological culture, appear to have played the same role in its formation. One of the most important and significant contributions to its rise and development is that of the Alans of the Northern Caucasus. In the mid-8th century, the central authority of Khazaria moved a large group of Alans from their homeland to the northwestern borderland of the khaganate, next to the lands of the Slavs in the forest-steppe zone along the Donets and Oskol rivers. The settlers skillfully and efficiently implemented an effective military and economic basis for the Khazar rule over that region. This has contributed to intensive contact respects between the (Alan) inhabitants of Khazaria and the Slavs. The high level of farming, craft activities, and military organization became the basis for the Khazar domination over the neighboring (Eastern) Slavs throughout the 9th and the first half of the 10th century. Like most other early medieval polities, the economy of the Khaganate was based on agriculture.

Studying the agrarian economy of the Saltiv culture in the forest-steppe zone represents therefore the key to the understanding of the economic power, as well as an important premise for the study of the military and political organization of the Khaganate, at least on its northwestern border. A dissertation defended in the mid-1980s was the last attempt to put together the results of various investigations concerning the economy of the Saltiv culture in the forest-steppe zone.[12] Since then, no other scholar have dealt with this research topic.

During the last decade of the 20th century and at the beginning of the 21st century, there was a surge of scholarly interest in the history and archaeology of the Khazars. Two congresses (one held in Jerusalem in 1998, the other in Moscow, in 2003) and a symposium (Kharkiv, 2003) took place at that time. To those, one can add the 12th International Conference on International Relations in the Black Sea Region (Rostov-on-Don, 2007), as well as the conference of Slavic-nomadic relations, which took place in Voronezh one year later (2008). Moreover, Khazar problems were discussed at the 7th International Conference on Jewish Studies (Moscow, 2010). Mass media in Eastern Europe suddenly began highlighting Khazarian history and archaeology. In both Ukraine and Russia, several dissertations and books were dedicated to this topic over the last two decades or so.

During that same period of time, the systematic study of archaeological sites attributed to the Saltiv culture in the forest-steppe area of Ukraine intensified. This is true for both settlement and cemetery sites. Not all of these materials are of equal value for our study of agriculture. To be sure, most important for research on the economy of ancient societies, especially at household level, are settlement sites. It is important to note in this respect that excavations resumed at Verkhnyi Saltiv over the past two decades. The strongholds and adjacent settlements at Mokhnach and Koropovi Khutory, as well as the settlement at P'iatnyts'ke I have also been the object of intensive study.[13] Several publications, including two books, put this considerable amount of new information to use.[14] However, even more archaeological, palaeobotanical and zooarchaeological data became available over the last few years, which can considerably enrich our understanding of the agricultural practices of the forest-steppe zone of Khazaria.

The material relevant to the topic of this book are currently stored in various institutions and organizations. For example, the materials from Verkhnyi Saltiv are now in the collections of the Historical and Archaeological Museum Verkhnyi Saltiv and the Archaeological Museum of the V. N. Karazin Kharkiv National University (KhNU). We have also looked and verified materials in the Institute of Archaeology in Kiev (for Verkhnyi Saltiv) and the "H. S. Skovoroda" National Pedagogical University in Kharkiv (KhNPU) (for Verkhnyi Saltiv, Mokhnach, Koropovi Khutory, and P'iatnyts'ke).

To obtain a full palaeobotanical spectrum from several of those sites, the pottery has been partially examined directly in the field during archaeological excavations (for seeds caught in the paste). At the same time, species were identified in palaeobotanical samples obtained by flotation. We have added the palaeobotanical samples from the hillfort excavated in Chuhuiv (materials now in the Repin House and Memorial Museum in Chuhuiv).

In short, most data used in this book have been obtained during the last two decades during the archaeological excavations carried out by the Medieval Archaeological expedition of "H. S. Skovoroda" State Pedagogical University (KhSPU, now KhNPU) in Kharkiv, led by Volodymyr Koloda. However, we also took into consideration the data known to us at the time the two monographs mentioned above.

With his book, Mikheev has set high standards for research on the early medieval agriculture. He dealt separately with the structure, typology, and technology of

12 Mikheev 1985.

13 Several settlement sites around the Mokhnach hillfort have been designated with letters of the Cyrillic alphabet. In this book, we opted for the corresponding letters of the Latin alphabet. Settlement P is therefore that designated as П in Ukrainian and Russian literature.

14 Koloda and Gorbanenko 2010; Gorbanenko and Koloda 2013.

production tools, mainly for plant growing. Unlike him, however, we refrained from generalizations across the entire territory of the Saltiv culture based on the material from a few sites. Our goal is to introduce new data against the background of the current state of research on the Donets region, and to advance their analysis on the basis of archaeometry. This goal is reflected in the structure of the book. Its sections are dedicated to specific interrelated problems: an overview of the history of archaeological research in the region with a focus on the most important sites; the presentation of the methods of analysis and the interpretation framework; a description, as detailed as possible, of the environment as one of the major factors involved in agricultural production. We envisage an analysis of site location based on various types of soils (employing the modern nomenclature and a simplified typology), and the position of the sites in relation to each other was taken into account. We have therefore attempted to reconstruct the agricultural practices of the Saltiv culture, as well as its animal husbandry. On that basis, we have drawn comparisons with the neighboring sites of the Volyntseve, Romny, and Borshchevo cultures attributed to the Slavs.

CHAPTER 1

Historiography and History of Research

In the southern parts of Eastern Europe, agriculture was the subsistence economy for a very long period of time, from Eneolithic and Bronze Age to the first half of the 20th century. Standards of living, various kinds of social relations, as well as domestic and foreign affairs of several states that existed in this area at different times — all depended upon it. The interest that historians and archaeologists have shown in the study of agricultural practices is therefore justified. Through their excavations, archaeologists can obtain direct information about the relevant tools and the biological remains of agricultural activity. What forms of farming were practiced depended upon climate and environmental conditions, but also on the level of manufacturing, in other words on how humans use natural resources to satisfy their food needs, including the organization of various forms of production, a quantity and quality of agricultural work tools, and an agrotechnical experience.

Despite the fact that the Khazar have been a topic of scholarly interest for almost 250 years, during the last third of the 20th century, the economic problems of Khazar Khaganate hardly attracted any scholarly attention. That is most likely because the written sources, upon which research was until then primarily based, provide very limited evidence on the economy of Khazaria. Moreover, the beginning of Khazar archaeology had its own problems. To be sure, the study of the Saltiv archaeological culture (as reflection of the material and spiritual culture of the many peoples of the Khazar Khaganate) already has a century-long history, during the first half of the 20th century. Archaeologists were primarily concerned with cemetery sites.

The excavation of settlement sites began first in the Don region just before World War II.[1] They could have provided sufficient material for an incipient understand of the economic development of Khaganate, but the publication of the excavation result was delayed for many years after the war. In addition, in the 1940s and in the early 1950s, the Soviet political regime had a hostile attitude towards the archaeological study and the publication of materials related to the Khazars.[2] A revival of Khazar archaeology cannot be dated before 1962, the year in which Mikhail Artamonov's remarkable monograph was published.[3] This book brought back Khazar issues in Soviet historiography, given that Douglas M. Dunlop's work was virtually inaccessible to East European scholars.[4] Khazar-age settlement sites became the focus of many archaeological excavations in the 1960s, and this first opened the possibility of studying the economy of certain regions of Khazaria on the basis of the archaeological data. Indeed, the first publications of the results of excavations of settlement sites greatly enriched the source base for the study of the economical life of Saltiv communities, especially of their agricultural activities.[5] Only five years after Artamonov's book, the newly discovered material became the basis for one of the most important Russian-language books about the development of the Khazar Khaganate that paid a great deal of attention to changes in economy.[6] However, the number of archaeological studies of the Khazar economy remained very small.

The situation began to change only in the 1980s. In Western Europe and in the United States, most scholars avoided any discussion of economy, as they were completely isolated from a direct study of the archaeological sources pertaining to Khazar history. A notable exception is the remarkable dissertation of Dieter Ludwig, dedicated to the social structure and the economy of the Khaganate.[7] Nonetheless, Ludwig's work is also based only on written sources. Some archaeological, but especially numismatic sources began to figure prominently in the studies that the American historian Thomas S. Noonan published in the journal *Archivum Eurasiae Medii Aevi* on the early medieval economy in Eastern Europe.[8] Specific issues, such as arable farming, animal husbandry, and crafts were almost entirely disregarded in those studies in question.

As all sites attributed to the Saltiv culture are in modern Russia and in the Ukraine, only Soviet scholars could provide a new impetus to the study of the economic history of the Khaganate. However, among those interested in Khazars, very few had an eye for economy. Perhaps the

1 Artamonov 1935 and 1936.

2 Pletneva 1990. It is important to remember that the territory of the Khazar Khagante was completely within the borders of the Soviet Union. The (independent) study of Saltiv sites was impossible for foreign scholars.

3 Artamonov 1962.

4 Dunlop 1954.

5 Liapushkin 1958a.

6 Pletneva 1967.

7 Ludwig 1982.

8 Noonan 1982, 1983, 1984, 1985, 1992b, and 1995–1997.

 | DOI:10.1163/9789004429574_003

most prominent examples of such rare kind of scholar in the late 20th century is the Kharkiv-based archaeologist Volodymyr Mikheev. Beginning with the 1960s, he excavated along the Donets a great number of sites of different kinds, but primarily settlements. This gave him a unique opportunity to accumulate and partially analyze a very rich and varied material. As a result, he wrote two dissertations, and published a on the social and economic development of the Don region of the Khazar Khaganate.[9] His book, which is the most accessible to scholars, dealt for the first time in detail with such matters as arable farming, animal husbandry, and basic non-food production (ferrous metallurgy, blacksmithing, pottery, wood-, leather-, and bone-working) of the Khazar population. Mikheev thus became a pioneer in the study of agriculture within the Khazar Khaganate.

Besides Mikheev's work, a number of general works on the archaeology of the Northern Caucasus region or of Crimea made references to, or even described briefly the development of agriculture.[10]

A new generation of archaeologists expanded even more the research interest in the Saltiv culture, especially in the Ukraine. Over the last two decades, the number of sites attributed to that culture that have been explored archaeologically has substantially increased. This is particularly true for hillforts and settlements, although the latter have until now attracted less scholarly attention. As a result, dozens of assemblages have been published. All of them formed the basis of two books summarizing the results of research on agriculture in the Saltiv culture. The first of them was devoted to crop cultivation and animal husbandry in the forest-steppe zone.[11] The other is an extended comparative analysis of the development of agriculture among the Slavs and within the Khazar Khaganate in a contact zone of the forest-steppe belt.[12] In addition, sites in the steppe region of the Khaganate are also taken into consideration. In recent years, the research on Saltiv settlement sites continued unabated. This made possible the accumulation of a significant body of evidence pertaining to the agriculture practiced in Khazaria.

In recent years, a number of zooarchaeological studies have been published, that dealt with the Saltiv culture in the forest-steppe zone of Khazaria.[13] A detailed analysis of the bone material has produced important conclusions regarding the role of animal husbandry in the northwestern region of the Khazar Khaganate. At the same time, such studies have revealed peculiarities for each site, a cautionary tale for the future against hasty generalizations.

The basis for this book is the body of materials accumulated over the last 20 years, largely as a result of the archaeological excavations of the Kharkiv National Pedagogical University, headed by one of the authors, on sites located within the Kharkiv region of Ukraine. To a large extent data from adjacent regions have also been employed. On one hand, we relied on "traditional" archaeological sources, on the other on the results of natural science investigations. "Traditional" archaeological sources include artifacts and features, and the locations of sites. Natural history materials derived from the analysis of palaeobotanical, zooarchaeological and palaeopedological data, in each case performed by specialists in the respective fields. We have also used data on the ancient climate. The synthesis of all those materials made it possible to obtain qualitatively new conclusions about the level of agriculture, in its quantitative and qualitative manifestations, as well as its particular traits in different regions. New data and a wide range of time-tested traditional and new methods have allowed the highlighting of the development of agriculture in the Khazar Khaganate.

1 The Research Area and Its Ethnic and Cultural Characteristics

The area of the forest-steppe zone between the Dnieper and the Don rivers is quite large 500 km long, 50 to 150 km wide.[14] As a rule, intense interethnic contacts took place on the southern boundary of the forest-steppe zone, while to the north, such contacts were more indirect. It is therefore important to establish the southern boundary of the forest-steppe zone; from east to west it goes from the Don river to the headwaters of the Oskol river, farther through the headwaters of Vovcha river, the Donets valley between 10 and 15 km to the south of the town of Zmiiv, then to the upper course of the Bereka river, the middle course of the Berestova river, and the lower course of the Vorskla, and finally to the Dnieper rivers[15] (map 1.1). The basin of

9 Mikheev 1968, 1985, and 1986b.

10 Baranov 1990.

11 Koloda and Gorbanenko 2010.

12 Gorbanenko and Koloda 2013. The analysis takes into account new data not included in previous monographs.

13 Koloda, Kroitor 2015, 2016 and 2017.

14 While the border between the steppe and the forest-steppe belts can be determined at least approximately (based on the presence or absence either of the steppe and or of the forest), the northern boundary of the forest-steppe belt is very difficult to establish, mainly because of the lack of clear criteria.

15 Different scholars prefer one of two versions of delineation in the east: from the Don river along the right bank of the Tikhaiaya Sosna, and then along the left bank of the Valui to the Oskol river; or along the right bank of the Chernaia Kalitva to the left

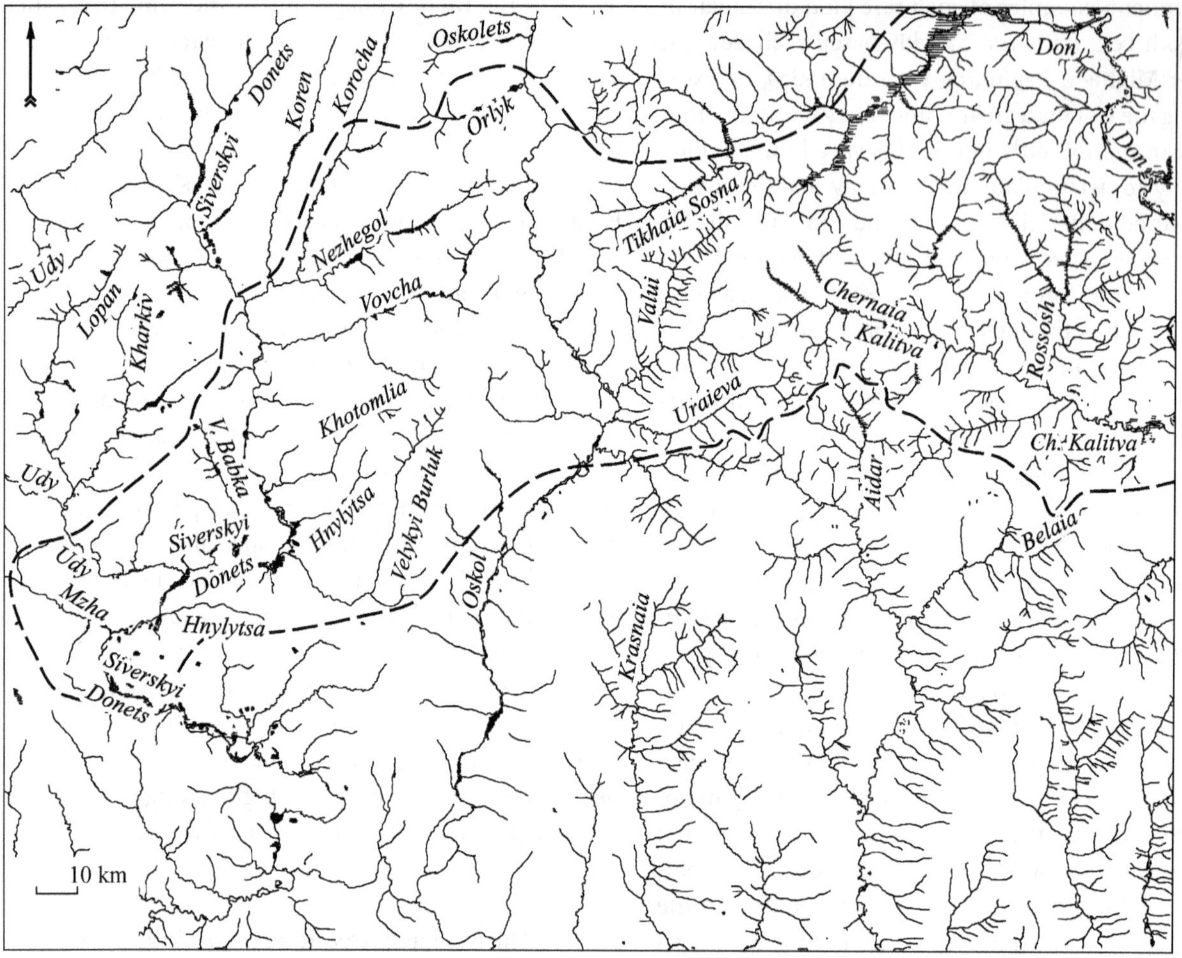

MAP 1.1 Map of the region under study

the Donets, especially starting from the upper Oskol River, was an area of very intense inter-ethnic contact.[16]

The archaeological investigation of the area between the Dnieper and the Don rivers has a long, century-long history. The mapping of sites dated to the last quarter of the first millennium, and attributed either to the Saltiv archaeological culture or to its neighbors, shows the degree of resolution achieved over the last few decades.[17] No significant changes in the distribution of sites may be expected in nearest future (map 1.2).

2 History of Research in the Region

The large area between the Dnieper and the Don has not been studied uniformly. The largest number of sites is within the territory of modern region of Kharkiv (Ukraine) and of the adjacent region of Belgorod oblast (Russia). The study of this restricted area has been going on for over a century.

During all that time, both field surveys and excavations have been carried out, which have made it possible to turn to specific issues of everyday life in the history of the ancient populations in the region. Given that a number of studies have already dealt with the history of research on Saltiv assemblages in the area, it is not necessary to go here into the details of historiography.[18] It is nonethe-

bank of the Uraieva, then to the headwaters of the Oskol river (15 km south from the confluence with the Valui). It should be noted that the former version overlaps the northern boundary of the distribution of Saltiv sites in the Don region, while the latter version corresponds well with the southern boundary of the forest-steppe variant of the Saltiv culture.

16 Another similar area was the southern border of the forest-steppe belt on the Don, where right-hand tributaries such as the Devitsa, the Potudan and the Tikhaia Sosna rivers formed the approximate boundary between the Khazar Khaganate and the Slavs. The Tikhaia Sosna region, together with the neighboring strongholds, is now generally recognized as the northern border of the Khazar Khaganate.

17 See, for example, Afanas'ev 1987, 10–20 and 168–184; Vinnikov 1995, fig. 1, 2; Kuza 1996; Kuchera and Motsia 1999.

18 E.g., Pletneva 1967, 3–12; Pletneva 1990; Mikheev 1985; Berezovets et al. 1986, 212–215.

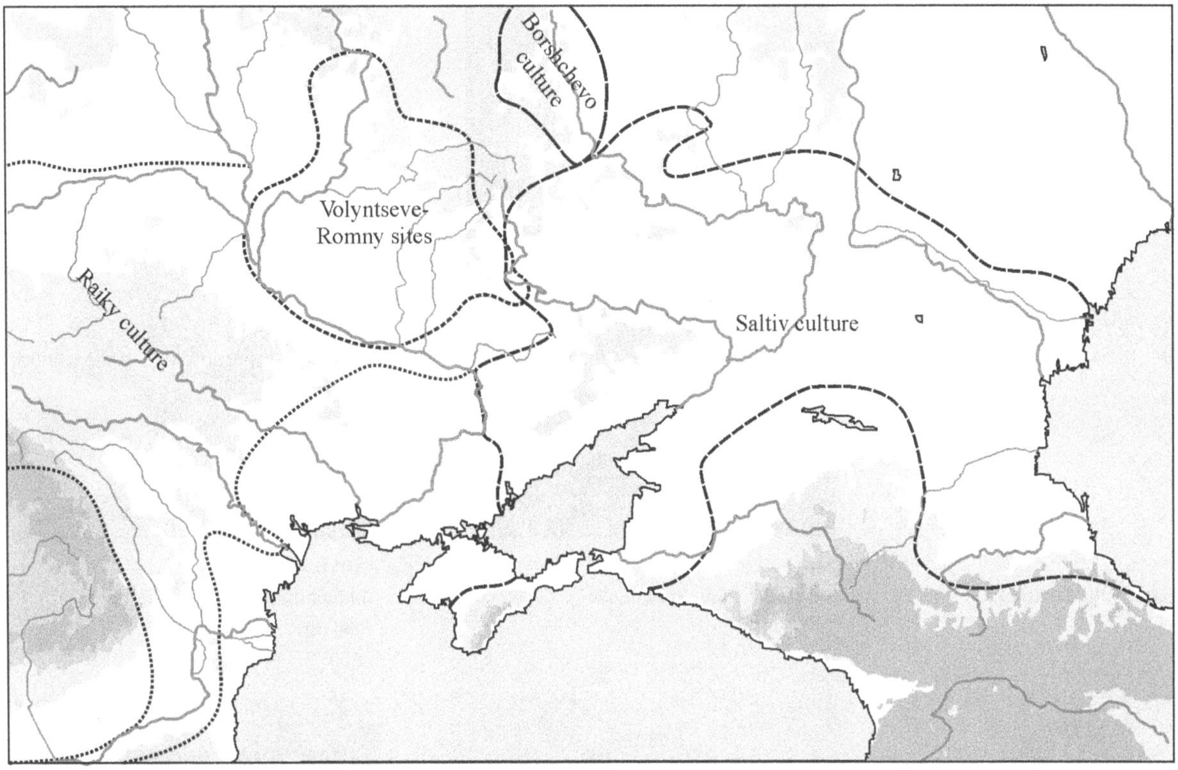

MAP 1.2 Distribution of the archaeological cultures in the southern parts of Eastern Europe in the last quarter of the first millennium

less important to mention a few outstanding scholars and their excavations, without which our own work would not be possible today.

The study Saltiv sites in the forest-steppe zone began ca. 1900, with the first excavations in the Mayaki hillfort (at the confluence of the Tikhaia Sosna and the Don rivers, in the modern region of Voronezh, Russia) and the discovery of the first catacombs near the village of Verkhnyi Saltiv (on right bank of the Donets, in the region of Kharkiv, Ukraine). One of the first archaeologists to deal with Saltiv culture was M. O. Makarenko (photo 1.1).[19] He began the excavations at Mayaki, both in the hillfort and in the open settlement. At the same time, an amateur named V. A. Babenko (photo 1.2)[20] began the excavations in Verkhnyi Saltiv. As a consequence, the combined names of the two sites (Saltovo-Mayaki) was chosen as a designation for the newly discovered culture.[21]

A significant growth in scholarly interest in Khazar-age assemblages in the Don region, in general, coincides with the first large-scale field surveys and excavations carried out by Mikhail I. Artamonov (photo 1.3).[22] After the discovery of the new culture, a new phase in its systematic study began in the late 1920s and is directly linked to Artamonov's activity.

Initially, more attention was paid to cemeteries, while settlements were the target only of surveys and trial excavations. The first excavations of the settlement in Verkhnyi Saltiv were carried out by S. A. Semenov-Zuser (photo 1.4) in the late 1940s.

The name of D. T. Berezovets (photo 1.5) is often associated with research on Slavic and Rus' assemblages. However, his contribution to the study of Saltiv sites is equally significant. Between 1959 and 1961, under his leadership, the Institute of Archaeology in Kiev carried out excavations at Verkhnyi Saltiv. Those excavations, which were done in anticipation of the Pechenega Reservoir being built on the Donets river, were a very important episode in the study of research on the Saltiv culture in Ukraine.

Ever since the late 1950s, on the other hand, the Scythian-Slavic expedition of "Maksim Gorkii" State University in Kharkiv (KhSU, now the V. N. Karazin KhNU) was led by B. A. Shramko (photo 1.6). Despite its focus, the expedition had a significant role in the study

19 Makarenko 1992.
20 Kadeeev 1996, 144–146; Chernigova 2000, 116–120.
21 In Ukraine, the simplified phrase "Saltiv culture" is currently employed. Saltiv is the Ukrainian name of Saltovo.
22 Medvedenko 2006.

PHOTO 1.1
Mykola Omelianovych Makarenko
(1877–1938)

PHOTO 1.2
Vasyl Oleksoiivych Babenko
(1877–1955)

PHOTO 1.3
Mikhail Illarionovich Artamonov
(1889–1972)

PHOTO 1.4
Semen Anatoliiovich Semenov-Zuser
(1887–1951)

PHOTO 1.5
Dmytro Tarasovych Berezovets
(1910–1970)

PHOTO 1.6
Borys Andriiovych Shramko
(1921–2012)

of archaeological sites (including that of Saltiv) located in the basin of the Donets.

By 1970, the archaeological research on the Saltiv culture entered another phase. At that time, Svetlana A. Pletneva (photo 1.7) began to explore a number of settlement sites of the forest-steppe zone along the Don. She headed the excavations in Dmitrievka between 1957 and 1973, and led the work of the international archaeological expedition at Mayaki (1975–1978). One unit of that expedition was headed by a prominent archaeologist, most famous for his studies of the Borshchevo culture, Anatolii Z. Vinnikov (photo 1.8). Beginning with the late 1960s, he began a thorough and very detailed study of assemblages in the forest-steppe region along the Don, and published several papers on Saltiv sites in that region, either alone or together with Pletneva.[23] Equally significant was the activity of another member of Pletneva's team, Gennadii E. Afanas'ev (photo 1.9). His work on the population of the forest-steppe area along the Middle Don at the end of the first millennium is still of great relevance to anyone interested in Khazaria.[24]

23 Vinnikov and Pletneva 1998.
24 Afanas'ev 1987.

PHOTO 1.7
Svetlana Aleksandrovna Pletneva
(1926–2008)

PHOTO 1.8
Anatolii Zakharovich Vinnikov
(b. 1940)

PHOTO 1.9
Gennadii Evgen'evich Afanas'ev
(b. 1946)

PHOTO 1.10
Volodymyr Kuzmich Mikheev
(1937–2008)

PHOTO 1.11
Volodymyr Vasyl'ovych Koloda
(b. 1955)

PHOTO 1.12
Gennadii Evgen'evich Svistun
(b. 1968)

Meanwhile, a student of B. A. Shramko, Volodymyr K. Mikheev, laid the foundations of the Kharkiv school of Khazar studies (photo 1.10). During the last third of the 20th century, he spent dozens of fieldwork seasons studying Saltiv assemblages in the Donets region. The excavations in that region are now organized by the "H. S. Skovoroda" KhNPU, and led by Volodymyr V. Koloda (photo 1.11). His students, Gennadii E. Svistun (photo 1.12) and Viktor I. Kvitkovskii (photo 1.13) continue the work on settlements of the forest-steppe area of Khazaria. Nataliia V. Chernigova (photo 1.14) from the V. N. Karazin KhNU has also excavated the settlement in Verkhnyi Saltiv. Moreover, a scholar from Luhans'k, Konstantin I. Krasilnikov (photo 1.15) has made a significant contribution to the study of Saltiv settlements in the steppe zone along the middle course of the Donets. The material resulting from his investigations offer us a great opportunity for comparison.

The early 21st century witnessed not only an expansion of research and publications on Khazar studies, but also an increased involvement of natural sciences in data analysis. This greatly enriched our understanding of the development of agriculture within the Saltiv culture, and its regional variations. A pedological and palaeoclimate study was conducted by a team headed by Iuryi G.

PHOTO 1.13
Viktor Igorevich Kvitkovs'kyi
(b. 1984)

PHOTO 1.14
Natalia Vasylivna Chernigova
(1971–2002)

PHOTO 1.15
Konstantin Ivanovich Krasil'nikov
(b. 1941)

PHOTO 1.16
Iurii Georgievich Chendev
(b. 1963)

PHOTO 1.17
Roman Vasil'evich Kroitor
(b. 1971)

PHOTO 1.18
Serhiy Anatol'iovych Gorbanenko
(b. 1978)

Chendev (photo 1.16). Roman V. Kroitor (photo 1.17) analyzed the faunal remains, while a large-scale gathering and processing of palaeobotanical materials was carried out by Serhiy A. Gorbanenko (photo 1.18).

The logic of the archaeological research in the general context of the historical study of populations of particular area and of different ethnic background requires first the attribution of new sites to the corresponding archaeological cultures, then the examination of that culture's territorial distribution, and finally the exploration of common issues of economic and social development. Only after that could more specific, local issues be investigated. One of those specific issues that concerns Khazaria is the development of agriculture in the northwestern borderlands.

3 Research of Basic Sites

The history of research on the archaeological sites that form the basis of our work, has been presented many times in the archaeological literature. Here we can only highlight some milestones. Most materials considered in this book have been presented in field reports with sufficient detail. This excavation documentation is kept in the archives of the Institute of Archaeology in Kiev and in the Institute of Archaeology in Moscow.

3.1 *Verkhnyi Saltiv*

The early medieval (Saltiv-culture) complex near the village of Verkhnyi Saltiv (in the Vovchansk district of the region of Kharkiv) has three components: a stronghold with a total area of about 6 hectares; 2) a large (open) settlement (more than 15 hectares large), which could be divided into a bailey and an extramural settlement at the foot of the stronghold; and four cemeteries with catacomb burials located on the right bank of the Donets River.[25] This complex includes burial assemblages with rich grave goods, which explains why it has always been the center of scholarly attention. The study of those cemeteries continues to this day, with some interruptions. A study of the early medieval settlement is slightly complicated by the fact that from the mid-17th century the modern village was built on top, in addition to a spa. While the archaeological exploration of the four cemeteries continued unabated since 1900, the settlement was first excavated in the mid-20th century, first the stronghold, and later the bailey and the open settlement In the mid-1940s, the State University in Kharkiv began excavations on the settlement site, under the leadership of S. A. Semenov-Zuser. Between 1959 and 1961, the excavations were organized by the Institute of Archaeology in Kiev, and led by D. T. Berezovets. The largest excavations, however, took

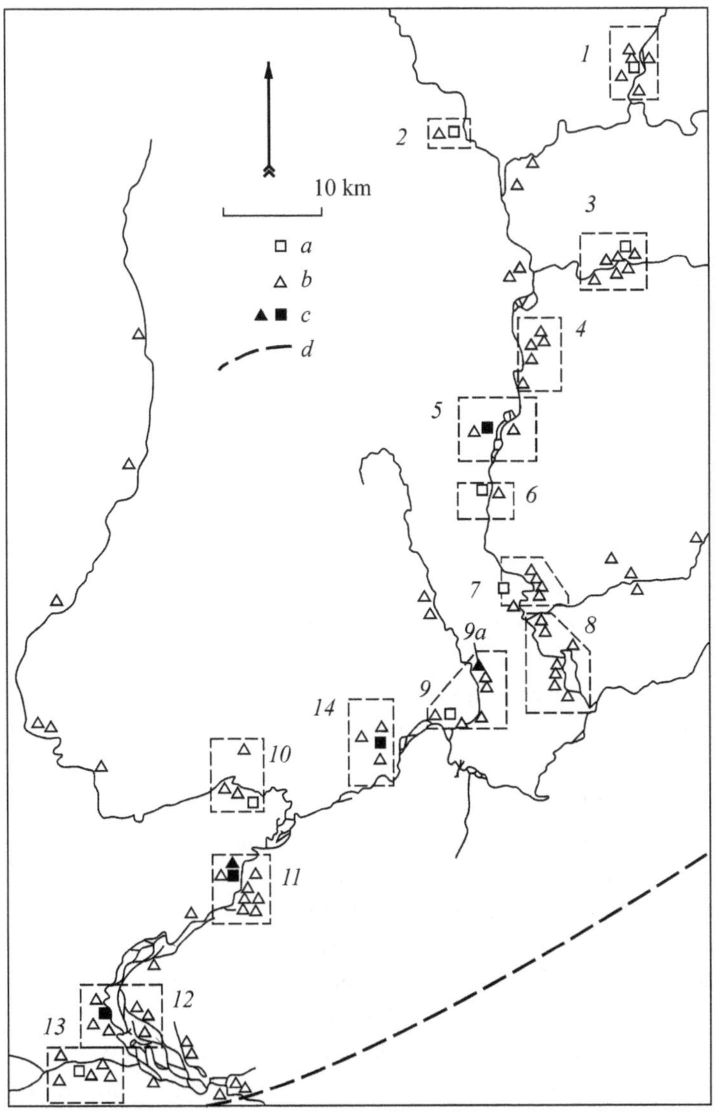

MAP 1.3
Sites of the forest-steppe variant of the Saltiv culture in the Donets River basin. Clusters of settlements: *1* — Dmitrievka cluster; *2* — Arkhangelskoe cluster; *3* — Vovchansk cluster; *4* — Grafs'ke cluster; *5* — ***Verkhnyi* Saltiv** cluster; *6* — Staryi Saltiv cluster; *7* — Khotomel cluster; *8* — Martova cluster; *9* — Kochetok cluster; (*9a* — **P'iatnyts'ke I** settlement); *10* — Kabanivs'ke cluster; *11* — **Mokhnach** cluster (hillfort and settlement P); *12* — **Koropovi Khutory** cluster; *13* — Sukha Gomil'sha cluster; *14* — Chuhuiv cluster. *Legend keys*: *a* — hillforts; *b* — settlements; *c* — the sites chosen for study; *d* — the border between steppe and forest-steppe.

25 A. V. Kryhanov's assumption about the existence of a fifth cemetery is not supported by the archaeological evidence. That the site at Netaylivka (settlement and burial ground), which is located on the opposite bank of the Donets, formed one and the same complex with Verknyi Saltiv site remains to be demonstrated.

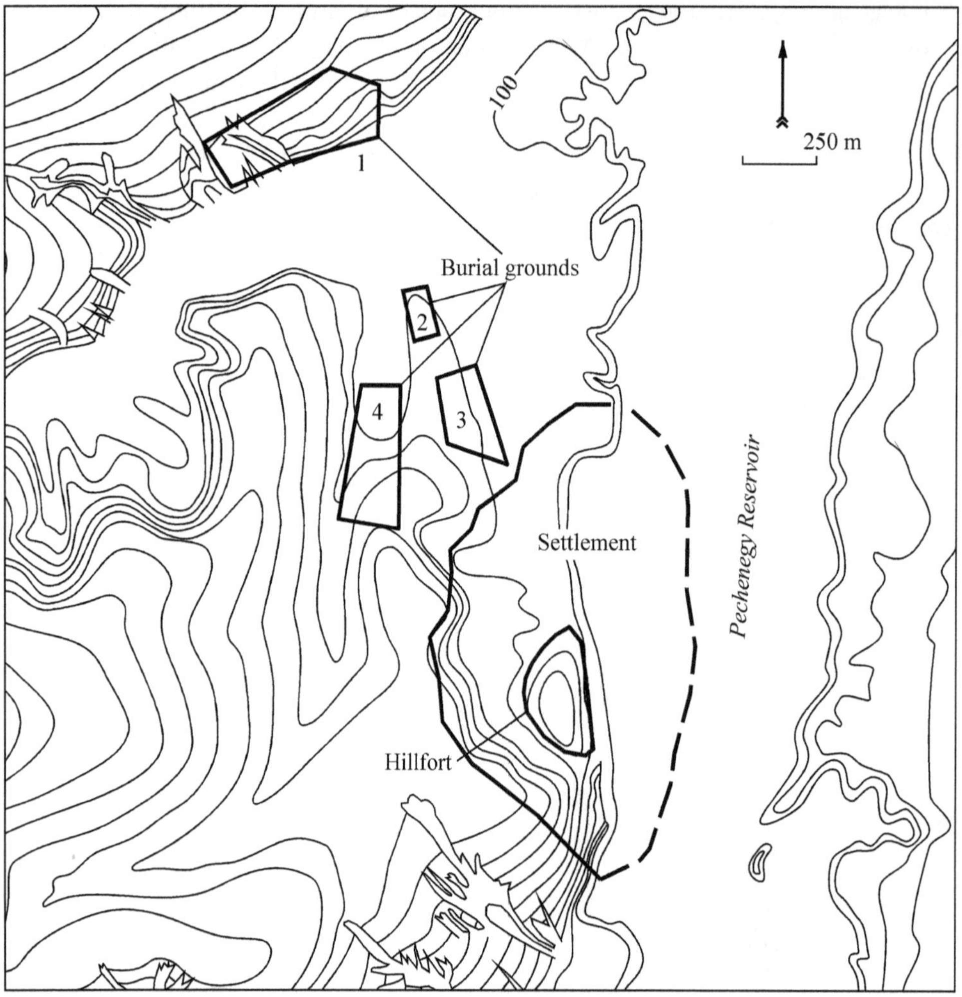

MAP 1.4
Plan of Verkhnyi Saltiv

place only between 1996 and 1998, then again in 2001 and 2004, and were led by Volodymyr V. Koloda from the Pedagogical University in Kharkiv. For several years (1998–2002), the stronghold was also explored by a team from the "V. N. Karazin" KhNU headed by N. V. Chernigova. As a result of all this activity, a very large area of the settlement has been already investigated: more than 2,000 square meters in the open settlement and about 1,000 square meters inside the stronghold.

A very large quantity of material, some without any parallel, resulted from those many years of excavation. The materials from Verkhnyi Saltiv play a key role in the arguments presented in this book. The zooarchaeological material was analyzed by V. I. Bibikova in 1959–1961 and then, more recently, by Roman V. Kroitor.

3.2 *P'iatnyts'ke I*

The settlement is situated on the western shore of the Pechenega Reservoir, not far from Kochetok. The site is in a wide floodplain, on the sand dunes on the left bank of the river Velyka Babka, a right-hand tributary of the Donets. P'iatnyts'ke 1 occupies a large area and stretches as a 1 km-wide strip for 2 km along the river (map 1.3; 1.5).

The site was discovered in 1957 by Svetlana Pletneva during a field survey. Excavations, however, started only two decades later, after the accidental discovery of a cremation burial with weapons. First B. A. Shramko, then Volodymyr Mikheev carried out small trial excavations, continued in 1988 and 1990 by A. V. Kryhanov. The latter opened up two other areas, one near the cremation burial, the other in the northern part of the settlement. In 2005, Mikheev also explored an area located to the northwest. Three years later, more excavations were carried out by the Pedagogical University in Kharkiv under the leadership of Volodymyr Koloda, while in 2009, P'iatnyts'ke was excavated by Viktor I. Kvitkovs'kyi, both in the north and in the south. The site was for the first time mapped out, along with the total area of 2,000 square meters so far dug (map 1.5).[26] Work continued to 2016,[27] and no less than

26 For more details on the excavation, see Kvitkovskii 2009. The largest area (949 square meters) was excavated in the north.

27 Personal communication from Viktor Kvitkovskii.

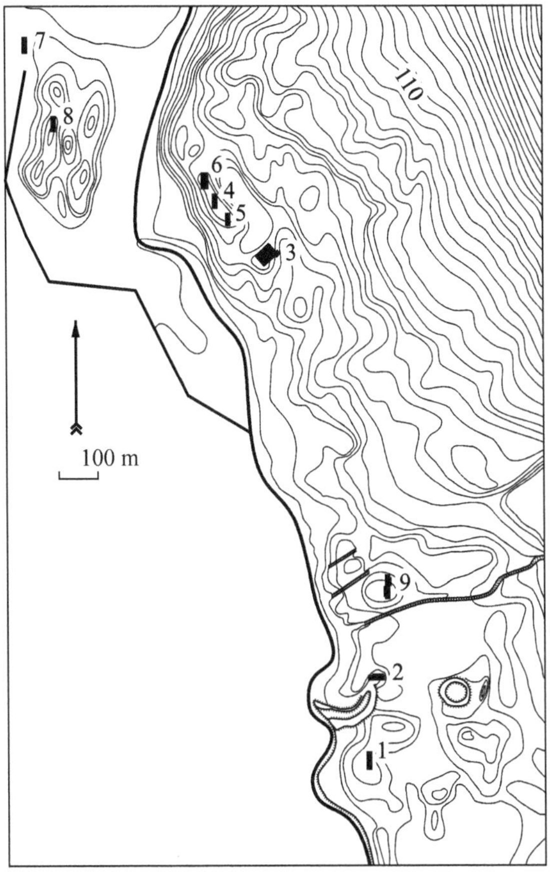

MAP 1.5
Plan of P'iatnyts'ke I with locations and numbers of trenches

100 settlement features have been recorded — 10 hourse , 10 silos, four workshops, and 93 pits of different functions. The large amount of archaeological information from the site makes it possible to study the agricultural practices of the settlement's inhabitants.[28]

3.3 *Mokhnach*

The settlement is located on an elongated, high promontory on the right bank of the Donets within the modern village of Mokhnach (in the district of Zmiiv, region of Kharkiv). There is a complex system of ramparts, which was associated with different building phases, as well as changes in population.[29] The site was indeed occupied during the early Iron Age (the forest-steppe variant of the Scythian culture of the 4th century BC), the early Middle Ages (the Saltiv and Romny archaeological cultures) and the modern times (second half of the 17th to the 19th century). During the early Middle Ages (Saltiv culture) the hillfort occupied an area of 12.5 hectares, and may have operated as an economic center of the north-western region of the Khazar Khaganate. As such, it was surrounded by numerous open settlements. No less than 17 such sites have been identified, the existence of which most likely coincided in time with the hillfort.[30] One of them was an industrial center, known as "settlement P" (map 1.3; 1.6).[31] Both the stronghold and "settlement P" have been studied in great detail, and because of that they constitute the basis for our analysis.

The stronghold is in fact known since the early 17th century. It is first mentioned in the Book to the Great Map of 1627.[32] V. V. Passek, from Kharkiv, first examined the site in the first half of the 19th century, but excavations only started during the second half of the following century. First, Boris A. Rybakov in 1950, then B. A. Shramko in 1953 carried out small, trial excavations inside the hillfort. By 1954, a small group headed by Svetlana A. Pletneva continued Shramko's excavations. Beginning with 199, with short interruptions, excavations were carried on the site by the Medieval Expedition of "H. G. Skovoroda" KhNPU headed by Volodymyr V. Koloda. By 2016, the area explored archaeologically was about 9000 square meters large (map 1.7).

The Saltiv assemblages were found in the northern part of the settlement located on the northern side of the stronghold. No less than 7,000 square meters have been explored archaeologically there, with almost 150 features, many of them associated with agriculture. The results of the excavations of assemblages attributed to the Saltiv culture were regularly published, although never in full, and with some delays. This only makes the site even more relevant for a study of agriculture in the context of the economy of the northern (forest-steppe) region of Khazaria.

The open settlement in Mokhnach was found in 2008 at a distance of 3.5 km to the north from the hillfort. Over three seasons (2009, 2011 and 2014), the Medieval Expedition of "H. S. Skovorod" KhNPU headed by Volodymyr V. Koloda opened up about 4000 square meters (map 1.8). Materials obtained during those excavations have a great significance for understating the level of development of soil cultivation and animal husbandry in the settlement, and its role in the agricultural development of the whole region with the Mokhnach hillfort at the center.

28 Kvitkovskii, Pashkevich, and Gorbanenko 2011a; Kvitkovskii, Pashkevich, Gorbanenko 2011b; Koloda, Kroitor 2015 and 2016.
29 Koloda 2005–2007 and 2007.
30 A total of 19 sites are currently known.
31 Koloda 2009b and 2010.
32 Serbina 1950, 71.

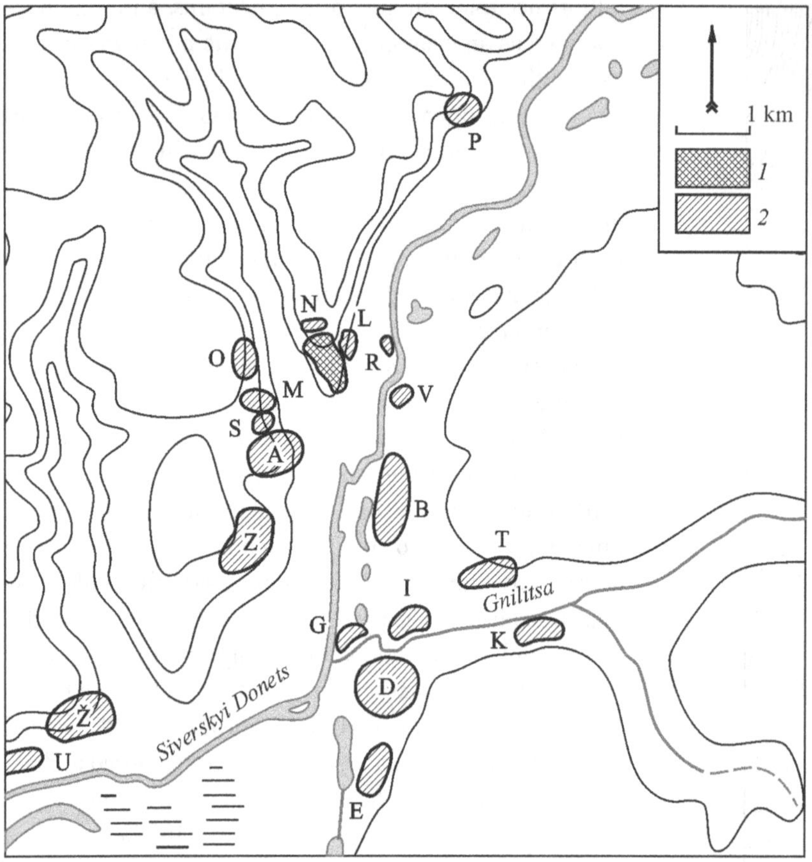

MAP 1.6
The microregion of Mokhnach with archaeological sites: *1* — hillfort; *2* — settlements

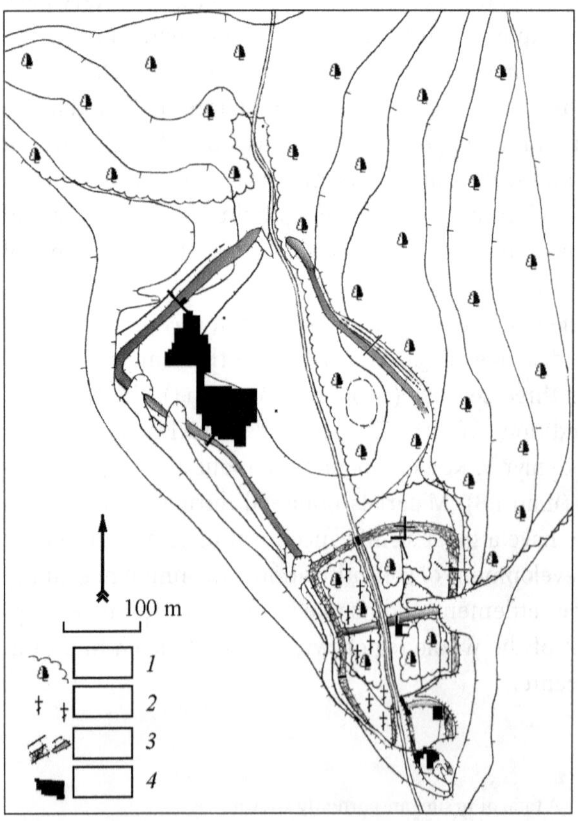

MAP 1.7
Plan of the Mokhnach hillfort: *1* — forest and its border; *2* — modern cemetery; *3* — ditches and shafts, ground, and masonry construction; *4* — digs

3.4 *Koropovi Khutory*[33]

The archaeological site at Koropovi Khutory, known to scholars for over a century, is located on the right bank of the Donets between the villages of Haidary and Koropove (in the Zmiiv district of the region of Kharkiv). Best known on the site are the hillfort located on the riverbank cliff (covering about 1.5 hectares) and a large settlement (about 30 hectares) occupying the slopes of a branched dell and the adjacent elevation to the west. The settlement surrounds the hillfort on the southwestern, western and northern sides (map 1.3; 1.9).

The territory of the site (both the hillfort and the settlement) is covered with depressions — remnants of sunken-floored buildings of early medieval date. Moreover, the territory of the settlement is terraced. The modern topography (hills, edges of ravines, and terraces) is marked with traces of ancient roads and trails, which crisscross both the hillfort and the settlement.

33 The place name has two spellings — Koropovi Khutory and Korobovi Khutory. The latter is more recent, and post-dates the beginning of the excavations on that site.

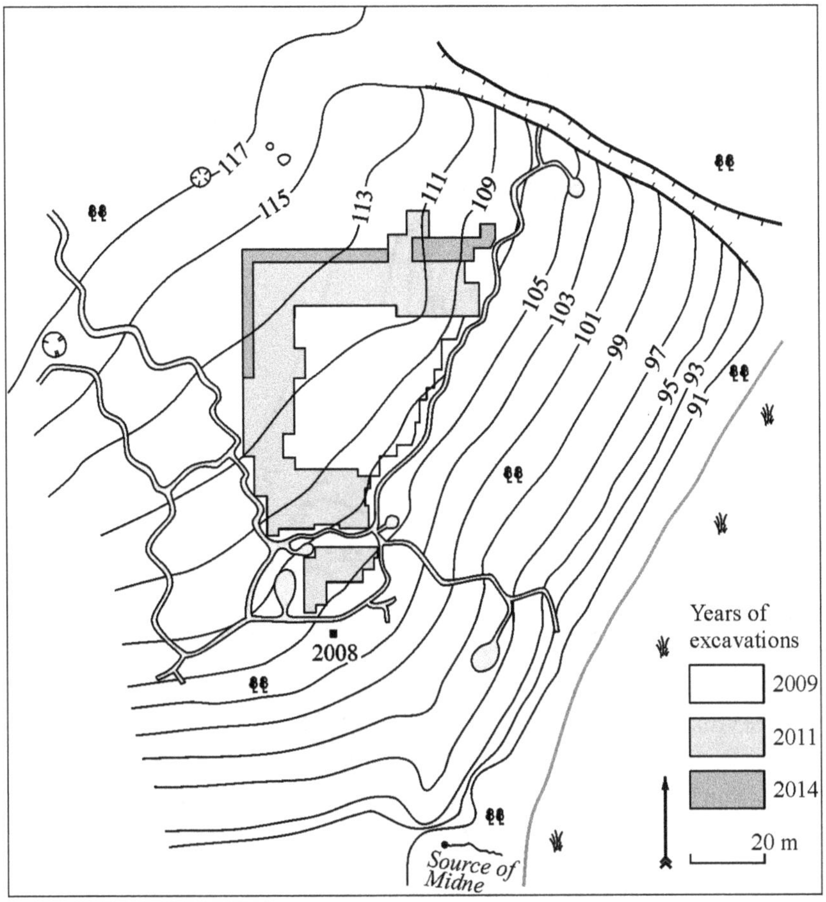

MAP 1.8
Plan of trenches on the Mokhnach P settlement site

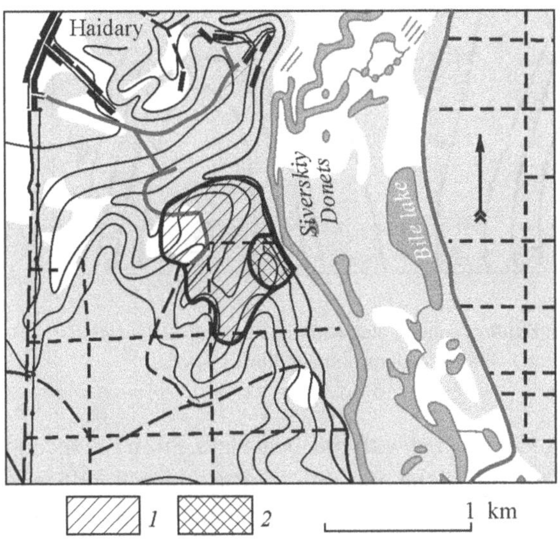

MAP 1.9
Koropovi Khutory: *1* — settlement; *2* — hillfort

The site was first mentioned in the early 20th century by D. I. Bagalei,[34] and first described by M. Fuchs in 1920.[35] After World War II (in 1953), B. A. Shramko undertook a first survey that revealed a multilayered settlement to the northwest from the hillfort, on the right bank of a stream.[36] The Saltiv features cut through a late Bronze-Age layer. Shramko carried out the first trial excavation inside the hillfort. Svetlana Pletneva visited the site in 1955 and noted the complexity of the local topography. However, she failed to mention the open settlement revealed by Shramko's survey. The latter's 1970 Scythian-Slavic expedition revealed another settlement at the foot of the former settlement on the western side. Excavations were carried out in this second settlement. Gennadii Afanas'ev visited the site in the mid-1980s, but he only inspected the rampart, which had been sectioned by Shramko in 1954.[37] Volodymyr Mikheev began excavations in 1998 and continued in the next year. His work focused on the hillfort, but he also cut a small portion of the settlement. Between 2003 and 2007, research on the site was conducted by the Medieval Expedition of the "H. S. Skovoroda" KhNPU under Volodymyr V. Koloda, who drew the plan of both hillfort and of (most of) the settlement (map 1.10). In addition, the expedition studied

34 Bagalei 1905, 35.
35 Fuks 1930, 104–105.

36 Wrongly indicated as located to the northeast from the hillfort in Koloda and Gorbanenko 2010, 19.
37 Afanas'ev 1987, 107–110 and fig. 64, 1.

the central part of the citadel, and the western rampart, in addition to yet another area of the settlement. The excavations revealed that the two settlements discovered by Shramko were in fact one and the same, large settlement, limited to the east and to the northeast by the Donets river, and to the north and to the west by a stream flowing from a natural lake 500 m to the west from the hillfort.[38] The southern part of the settlement is located in the upper reaches of a deep ravine that goes to the right bank of the Donets to the north of the hillfort, while the south-western part of the settlement is on a spur of the same ravine (another stream flows there, which dries up in the summer).

Over many years of excavations, about 2,000 square meters have been explored archaeologically, but that represents only about a fifth of the territory inside the ramparts of the hillfort, and less than 1 percent is the settlement area. Nonetheless, that is still sufficient for drawing some conclusions about the general history of the site, especially the settlement. There are three distinct phases of occupation: late 5th to 4th century BC (Scythian age); mid-8th to mid-10th century AD (Saltiv period); and second half of the 10th to the mid-11th century (Romny culture). In addition, artifacts have been found that can be attributed to the Gorodets (5th century BC to 7th century AD), Pen'kivka culture (7th to mid-8th century AD) and Slobozhanshchina culture (late 17th to 19th century). The defensive works (ramparts) have two main building phases, which coincide with the early Iron Age (Scythian period) and the early Middle Ages (Saltiv and, perhaps, Romny period).

The Iron-Age occupation phase concerns the hillfort primarily, while the Saltiv-period occupation is within a large area of the settlement (the Romny-period occupation was restricted again to the hillfort). In all phases of occupation, the subsistence economy was based on arable farming and animal husbandry, supplemented by fishing, hunting and handicrafts. One of the most important industrial activities metalworking, but there is also evidence of carpentry, bone- and antler-working, as well as spinning and weaving. Trade relations with the early medieval populations in the Crimea, the Sea of Azov region, and along the Volga have been documented archaeologically. The archaeological record dated to the end of the first and of the beginning of the second millennium confirms the idea that in forest-steppe region along the Donets, the multiethnic population to which the Saltiv culture is attributed coexisted with the local Slavs. From the second half of the 8th to the mid-10th century, the militarily and politically dominant factor in the northwestern region of Khazaria were the Alans. After their return to the North Caucasus region at some point during the second half of the 10th century, the dominant factor in the region is believed to have been the Slavic tribe of the Severians mentioned in the Russian Primary Chronicle. The multiethnic population to which the Saltiv culture is attributed was now included in their ethnic community, and most likely assimilated.

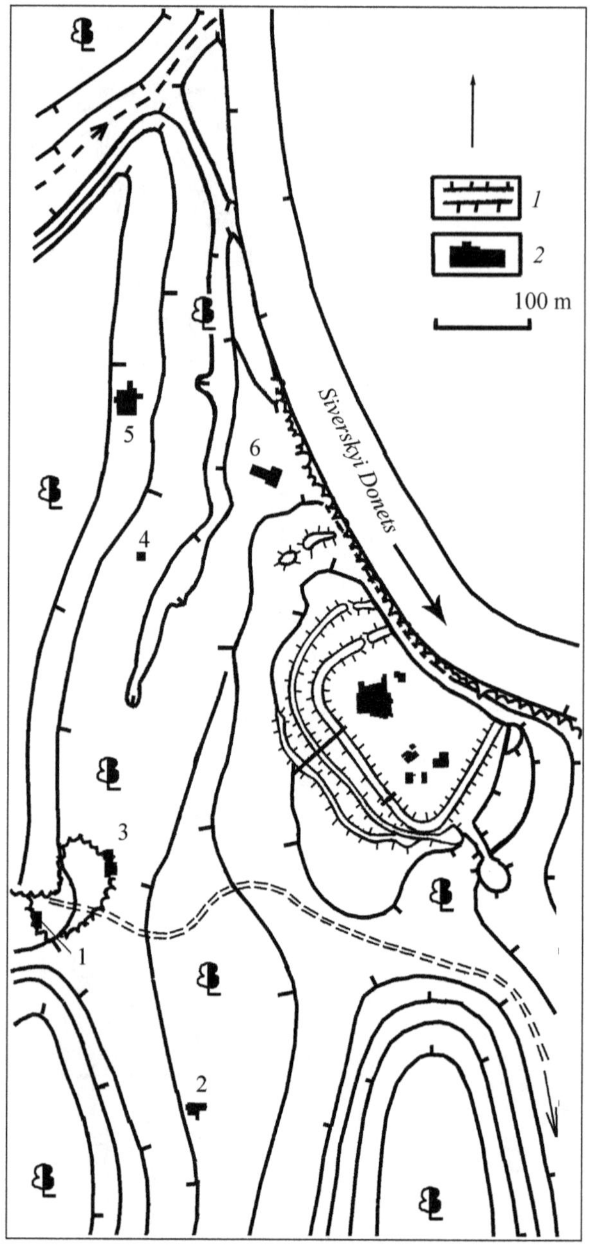

MAP 1.10
Koropovi Khutory, general plan with the numbers and location of trenches: *1* — fortifications of the hillfort; *2* — digs

38 Koloda and Gorbanenko 2010, 21 wrongly indicate the location as being to the east from the hillfort.

The excavations results have been published, with a more detailed discussion of both pottery and dwellings.[39]

The majority of the sites considered in this book are located in the basin of the Donets, directly in a contact zone. However, in order to bring to the fore key features of the discussion, we have also employed data from other settlements of the Saltiv archaeological culture in the forest-steppe and steppe zones of the Khazar Khaganate, as well from settlements of the neighboring Slavic population.

39 Koloda 2002–2003, 2005, and 2007. See also Kvitkovskyi 2008 and 2012; Koloda 2008 and 2009a.

CHAPTER 2

Methods of Interpreting the Agricultural Material

The ideas and methods of analysis and interpretation are not new.[1] They are now called environmental and landscape archaeology, and are concerned with the reconstruction of historical landscapes, natural conditions, and the interactions of humans with nature in the past. As a consequence, the main sources for such studies are those produced by natural sciences. Those methods began to develop in Western Europe over the last century, and still occupy a central position.[2] In Eastern Europe, such methods of research were first applied to the history of the southern regions during the second half of the first millennium. However, the results remained unknown until recently, as they have not been published in English.

Much like in previous studies, we will employ here all possible methods of research in order to reveal the details of the economic life of the population living in the forest-steppe zone between the Dnieper and the Don during the early Middle Ages.[3]

The historiography of the early medieval agriculture in Eastern Europe has been covered in a few specialized studies. Shortly after World War II, V. M. Slobodin surveyed the literature published before 1917.[4] Some results of the study of agriculture in the Soviet Union were discussed later by Iurii A. Krasnov.[5]

A presentation of the state of research in palaeobotany in Ukraine may be found in G. O. Pashkevich's work.[6] The current state of research has been recently summarized in two studies.[7] Because of that, the topic of this chapter is not the history of how interest in the early medieval agriculture developed, but what the interpretational potential is that various models of analysis have.

1 General Analysis of Agriculture (Fig. 2.1)

For the general study of agriculture, the minimal unit of analysis is the settlement. In other words, at stake is often a local level of research for clear characteristics of the relations between different components of agriculture. As such, one needs to take into consideration both movable and immovable archaeological materials, as well as the palaeoecological conditions. Although the availability of such data depends upon the degree of scrutiny involved in the investigation of the site, palaeoecological data have become widely available on recently excavated sites. In short, we selected sites for analysis on the basis of how much archaeological material was available.

The main components of agriculture are the anthropogenic and natural factors, but also their mutual influence and interaction. Among the natural components, the key factors are climate, topography and soils. During the first millennium AD the climate fluctuated slightly, but significant changes in relation to modern climate conditions. The climate optimum cannot be dated before AD 1000. Scholars studying the history of climate changes on the territory of Ukraine have also paid attention to the first millennium AD.[8] Since the reconstruction of the climate history is quite complete, and more recent studies, while adding details, did not contradict it (see Chapter 3), one can use the general charts of climate changes when analyzing from a particular region or site. Moreover, the general framework of climate history does not preclude the significance of the local landscape (relief).

The anthropogenic factors refer to the tools employed in agricultural production. From an archaeological point of view, the basic data in that respect results from details of tools for primary soil tillage with the use of draft animals — metal components of wooden tools, such as socketed ard-heads (plowshares) and coulters — or of tools employed with human power for primary or secondary tillage — functional parts of hoes with horizontal and vertical socket, or metal reinforcements for spade blades. Equally significant are tools for harvesting — sickles and scythes. A third category of archaeological evidence pertaining to the anthropogenic factors is the group of finds that includes implements for food processing — graters, mortars, and manual rotary millstones — as well as storage

1 The arguments presented in this chapter have been presented in greater details in Gorbanenko 2007; Gorbanenko, Zhuravl'ov, and Pashkevich 2008; Gorbanenko and Pashkevich 2010; Koloda and Gorbanenko 2010; Karavaiko and Gorbanenko 2012; Gorbanenko and Koloda 2013.

2 E.g., Chisholm 1968; Jahnkuhn 1977.

3 Koloda and Gorbanenko 2010; Gorbanenko and Koloda 2013.

4 Slobodin 1952, 9–14.

5 Krasnov 1969.

6 Pashkevich 2002 and 2005.

7 Gorbanenko 2007, 5–25; Gorbanenko and Pashkevich 2010, 7–34.

8 Bezusko and Klymanov 1987.

 | DOI:10.1163/9789004429574_004

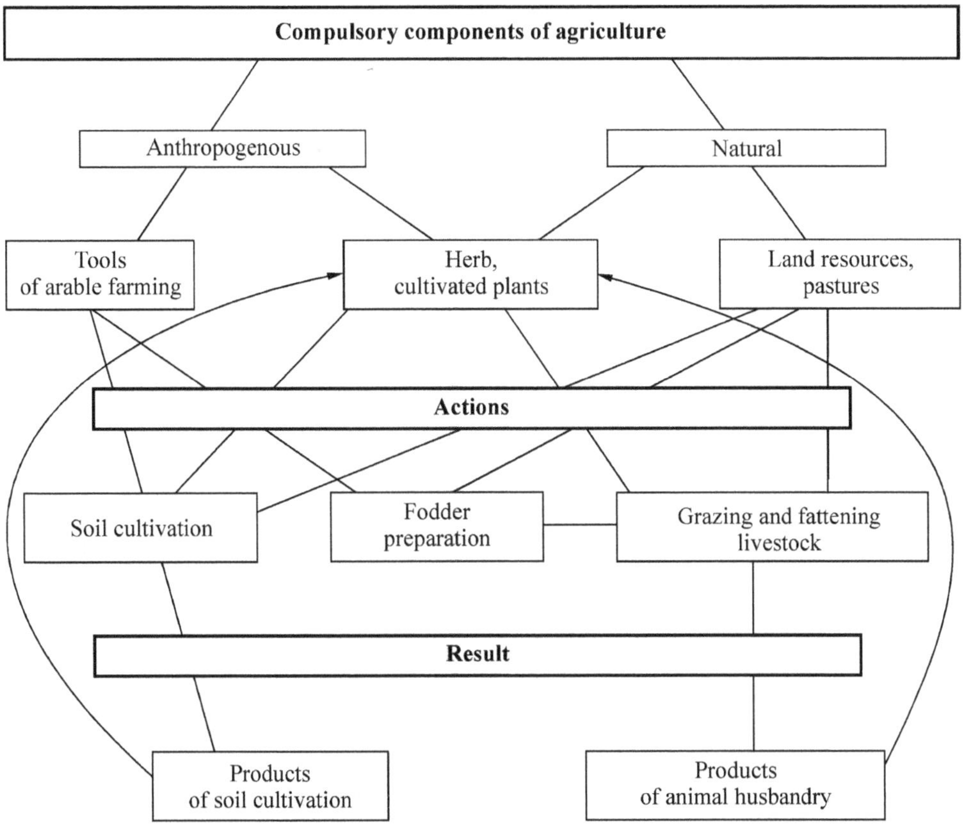

FIGURE 2.1
Interactions of agriculture elements

facilities — silos, barns, and earthenware storage vessels. Finally, anthropogenic factors also include tools for animal husbandry — cattle bells, shears, butcher knives, and harness components — as well as archaeological features that could be associated with livestock — stables, byres, and pens. The relation between crop cultivation and animal husbandry is relevant for the study of the interaction between the anthropogenic and natural factors.

Agriculture, therefore, involves both working the land (for crop production) and grazing the livestock. The link between the two aspects of agricultural production is the fodder prepared with scythes, as well as the manure (animal waste mixed with byproducts of soil cultivation, such as straw or waste products of threshing. As the main form of land work, tillage requires draft animals, such as oxen and horses. According Veniamin Tsalkin, draft animals typically represent about 5 percent of the entire livestock.[9] After harvest (using sickles and scythes), crops are processed by means of mortars (for cleaning of grain from husk and for grinding in order to produce groats), graters (to produce groats and flour in small amounts), and lightweight manual rotary millstones (to produce flour and occasionally groats). What is not processed is typically stored away in various containers and in silos, in order to keep seeds for the following year. What is left after harvesting and processing crops (straw and waste products) could be used as fodder. However, the most common type of fodder was hay stored over the winter.

Grazing the livestock typically involved access to pastures, and at time, transhumance.[10] Fields left fallow, those near the forest or not otherwise suitable for farming, as well as neighboring floodplains — all could serve for grazing needs. Palaeobotanical and zooarchaeological data can provide crucial information about the processing and storage of crops, or the use of the livestock (whether for meat or for dairy).

1.1 *Research of Settlement Resource Areas*

Archaeologists have become accustomed to the idea that the ideal catchment area of any settlement is a circle with a 5-km radius. Within that area, all key resources employed by the settlement inhabitants were used.[11] For this type of

9 Tsalkin 1969, 94.

10 Buniatian 1997 (we are grateful to the author for her clarification of the concepts of *pasturable grazing* and *transhumance*). Transhumance involves seasonal migrations over several hundred kilometers, following the herds from spring to autumn. Nonetheless, we employ the term here for grazing animals within a much shorter distance (a few kilometers) away from a settlement.

11 See Tomashevskyi 2010, 176–77.

analysis to work, one also needs to have a good knowledge of the zone's elevation (relief). Given that the relief has formed a long time ago, it is quite possible to predict resource areas on the basis of maps of the modern landscape configuration (at a scale of 1 : 100 000 or larger).[12] On that basis, one could distinguish places with soil types most suitable for crop cultivation, and places with sufficient water that were better used for animal husbandry.[13] Soils around settlements are analyzed on the basis of palaeopedological data. We have used here a rather simplified typology of toils.[14] Some soils were formed under the forest canopy, primarily through the decomposing of leaves and pine needles, and are typically used for swidden cultivation (slash-and-burn clearing). By contrast, chernozems formed through the decomposition of the steppe vegetation and are used for arable farming. The formation and fertility of some soils depended upon humidity (they may be found near riverbeds, in lowlands, as well as in marshlands); their fertility may be restored when the soil-forming elements penetrate through groundwater and sewage. Those are light soils for plowing, and do not typically need artificial measures for renewing fertility. Finally, there were soils unsuitable for cultivation, which were therefore not used for farming.[15] In our opinion, medieval farmers were capable of distinguishing between those four basic types of soils, which they used accordingly. The distribution of the four types of soil may therefore indicate the type of farming and the techniques employed to restore fertility.

The environment was further reconstructed on the basis of the assessment of vegetation based on the current data in the region or on the settlement site. While complementing the picture of the original form of farming based on the corresponding resource zone, this information makes it possible to allocate various plots of land to different agricultural purposes.

Along with the information on the environmental conditions, the analysis will use the archaeological information regarding the tools for soil cultivation (Fig. 2.2). Sometimes, the latter may be complemented with information from ethnographic reports (Fig. 2.3).

The latter is particularly important, when the archaeological information is restricted to details of the tools for soil tillage.[16] There are five possible devices, some with, others without parts surviving in the archaeological record (Fig. 2.4).

Bow ards with ard-heads without iron tips are positioned at an angle of about 45°. By contrast, body ards with ard-heads without iron tips are built to run horizontally, parallel to the ground. Bow ards with ard-heads with narrow shares are similarly positioned at the angle of about 45°, while body ards with ard-heads with broad shares run horizontally, parallel to the ground. Plows are typically body ards with ard-heads with broad shares placed horizontally, parallel to the ground, but additionally equipped with coulters and mouldboards.[17] It is important to note that the presence or absence of support wheels had no effect on the tillage technique, but considerably improved the ability of the farmer to control the implement.

On the basis of the analysis of currently known details of early medieval tools for soil tillage, but taking into account ethnographic data as well, we propose an outline of the development of arable tools over time (Fig. 2.5).

Improving different types of tools for tillage, the most common, simple types (including the plows meant to turn the turf slightly) were already in existence in the 3rd and 4th centuries, as confirmed by assemblages attributed to the Sântana de Mureş-Chernyakhov culture (Fig. 2.6, *5*).

During the subsequent centuries, the use of arable tools on different sites should be interpreted not in terms of different levels of development, but in terms of the adaptation to specific natural conditions in the region where the site was located. Three possible interpretations may therefore be advanced for the reconstructed tools for tillage. Where iron details are absent, bow or body ards may have been used that had no shares, especially in floodplains , where the soil was relatively easy to break. Alternatively, such tools were not needed because no farming was practiced in that particular locale. Where narrow shares appear, especially in the forest belt, this may well be an indication of swidden cultivation by means of slash-and-burn. Even where chernozems are present, finds of such shares are an indication of clearing of the previously forested area. Furthermore, this is an indication that the area in question was turned to arable farming. Finds of broad shares, often accompanied by coulters, indicate a high technological level of tillage, as well as intensive (prolonged) cultivation. When associated with forest soils,

12 See http://www.wikimapia.org (visit of February 29, 2020). A good example is a map of Ukraine at the scale indicated. One should always prefer maps created before the intensive anthropogenic intervention of the 20th century (e.g., through the buildings of reservoirs).

13 See Gorbanenko, Zhuravlev and Pashkevich 2008 for similar observations regarding the Pastyrs'ke hillfort.

14 Gorbanenko 2007, 32 and 33 fig. 4. The first to attempts a unification of soils with similar nomenclature was Tomashevskyi 2003, 149 (see also Tomshevskyi 2008).

15 See Vernander and Tiutiunnik 1986.

16 In this book, we employ Iurii Krasnov's typology of the iron parts of implements for tillage (Krasnov 1987).

17 Gorbanenko 2006, 74 fig. 1.

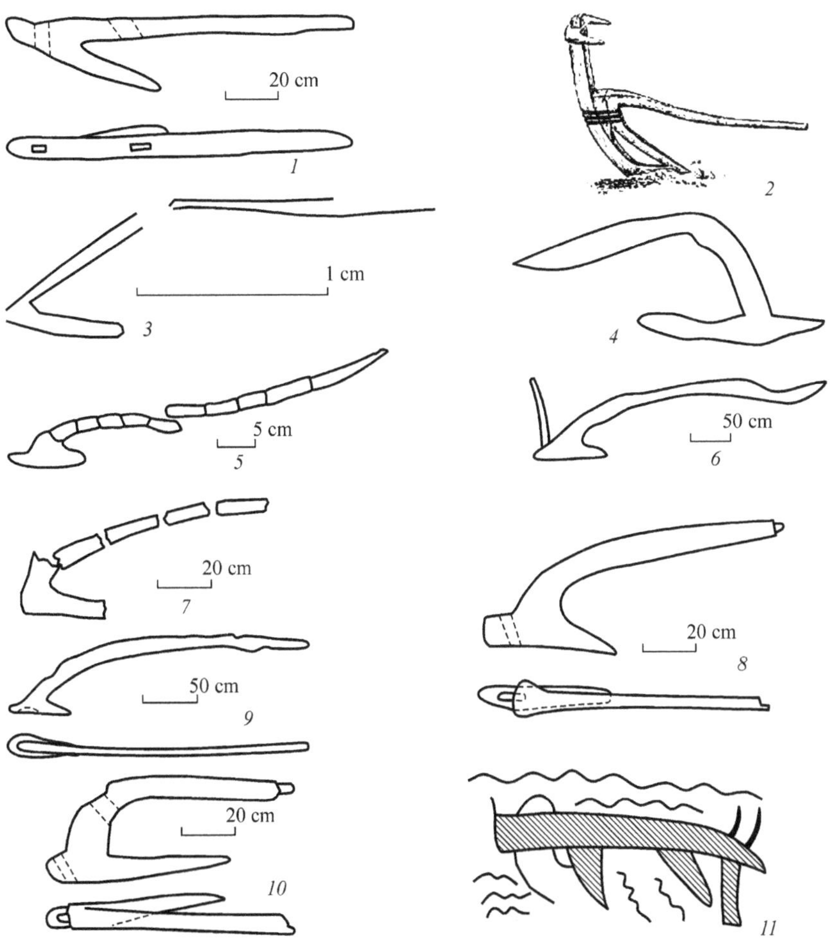

FIGURE 2.2
Tools for primary soil tillage with draft animal power known: *1* — Brest; *2* — Khrinnyky (reconstruction); *3* — Pisochnyi Riv (picture on a spindle whorl); *4* — Troits'ke (picture on a spindle whorl); *5* — Bil'sk (clay model); *6* — Polissia; *7* — Verkhnia Maivka; *8* — Serhiivs'ke peat-bog; *9* — Kaplanovychi; *10* — Tokarivs'ke peat bog; *11* — drawing on a pot of the Sântana de Mureş-Chernyakhov culture (Lepesivka) (*1* — after Korobushkina 1979, 16–17, fig. 1, *1*; Krasnov 1987; *2* — after Kozak, Pryshchepa, and Shkoropad 2004, 8–10, fig. 3–5; *3* — after Karavaiko and Gorbanenko 2012, fig. 3.2, *1*, 3.3, *1*; *4* — after Dubynin 1966, 270–272, fig. 1, *2*; *5* — after Shramko 1984, 255; Shramko 1996, fig. VIII, *10–12*; *6* after Shramko 1972, 30–31; *7* after Likhachev 1982, 51–53; *8* after Shramko 1964, 84–100; *9* after Krasnov 1987, 84–85; *10* after Berezovets 1952, 174–175; *11* after Rybakov 1962, 87; made by S. A. Gorbanenko)

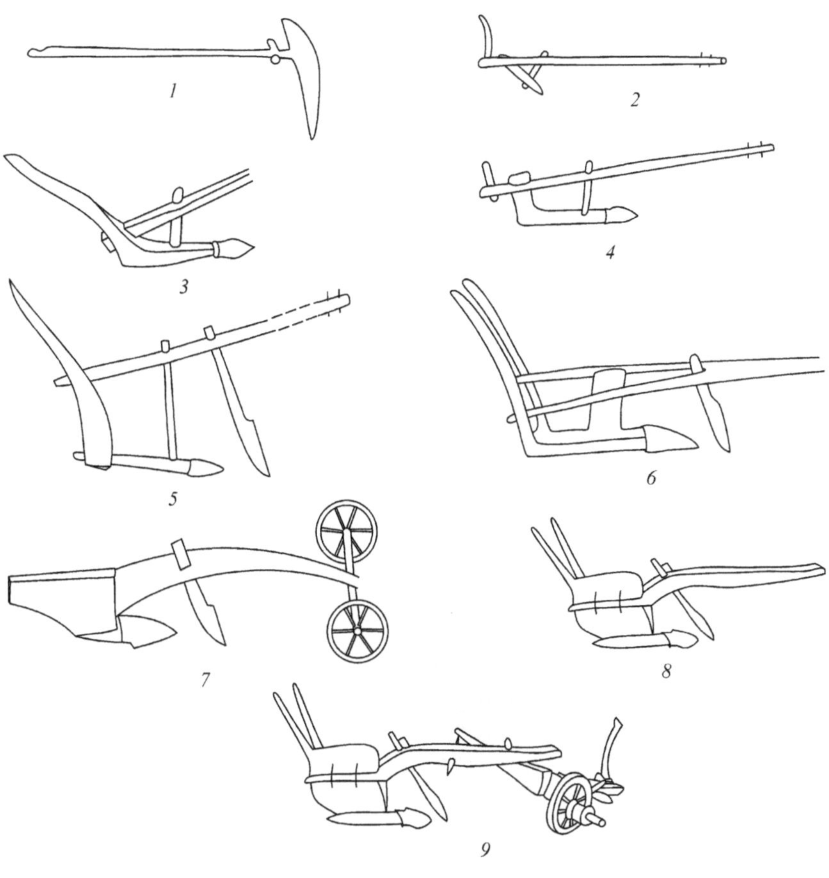

FIGURE 2.3
Tools for primary soil tillage with draft animal power (*1–9*), known from the ethnographic data (after Guldenstedt 1804, 3–5; Zelenin 1907, 12–18; Zelenin 1991; Mamonov 1952, 73–75, and others; made by S. A. Gorbanenko)

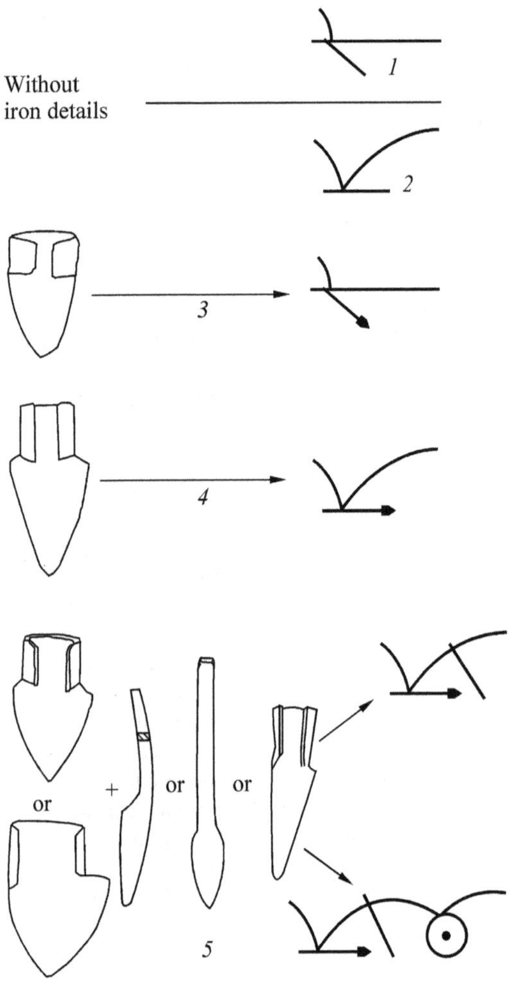

FIGURE 2.4
Iron details and tools for primary soil tillage with draft animal power (*1–5*)

this may well be an indication that swidden cultivation has long been abandoned in favor of fallow or other forms of arable farming.

Tools for tillage using human power may be classified on the basis of the socketed hoe-heads — with a vertical, disconnected socket, or with a horizontal solid socket. The former was a tool much easier to manufacture, while the production of the latter was more complex. The latter was closer to a modern hoe in terms of quality. Vertical sockets allowed the attachment of the iron part onto a curved shaft (Fig. 2.7, *1*). A hoe with a horizontal socket had to an ordinary, straight shaft (Fig. 2.7, *2*). Hoes were used for digging pits,[18] for cultivating small ("garden") plots,[19] for primary soil tillage,[20] or even for woodcarving, when used as adzes.[21] In addition to hoes, farmers in communities of the Saltiv culture employed spades with iron reinforcements for the blade. Some reinforcements are rectangular, others are semicircular (Fig. 2.8). Those spades were not suited for long-term sod grounds; they were instead used for cultivation of soft soils, either in floodplains, or in fields that have previously been plowed. Finds of such tools do not provide information for assessing the level of the agriculture. In terms of numbers, they may be an indirect indication of its development. Moreover, the use of hoes with horizontal sockets shows a sufficiently elevated level of iron processing.

When dealing palaeobotanical data, both charred seeds, grain imprints on pottery may be used. Charred seeds may be found in bulk, through direct excavation or by means of flotation (which may also reveal non-charred grains). Flotation has been used for the collection of palaeobotanical data since the 1960s and early 1970s.[22] Around that same time, the first methodological articles were published on using a foam solution based on water.[23] That was a revolutionary method that brought palaeobotanical research to a new level of development.[24]

Meanwhile, it has become obvious that random finds of grains, even in large amounts, cannot give a complete picture of the ratio of crops. In such cases, one or two species often prevail, while others are present in much smaller numbers, or altogether absent. For example, at Rohalyk (Luhans'k region), a Saltiv settlement in the steppe belt, a great number of charred seeds have been found, The samples were dominated by barley with a significant admixture of oats, while other cereals were rare or absent.[25] By contrast, using targeted flotation on a number of settlements in Obukhiv near Kiev, a much more statistically viable result was obtained.[26]

Another way for obtaining palaeobotanical data is by studying imprints of grains and seeds of cultivated plants on pottery. After examining the pottery from any given site, the fragments with imprints are transferred onto a plastic matter (commonly plasticine), before reproducing a model. The method was introduced in the USSR by the Moldovan palaeobotanist Zoia V. Ianushevich in the mid-1960s.[27] Until recently, all grain and seed imprints on pottery were identified by Galina O. Pashkevich, but after 2010, the task was reserved to Serhiy A. Gorbanenko. Any

18 Pletneva 1989, 91–93.
19 Mikheev 1985, 38; Magomedov 1987, 63.
20 Mikheev 1985, 38–39.
21 Koloda 2014c, 73 fig. 4, *12*.
22 Streuver 1968; Jarman, Legge, and Charles 1972.
23 Jarman, Legge, and Charles 1972.
24 Dennell 1978.
25 Pashkevich 1991c, 17.
26 Kravchenko, Pashkevich 1985.
27 Ianushevich and Markevich 1970.

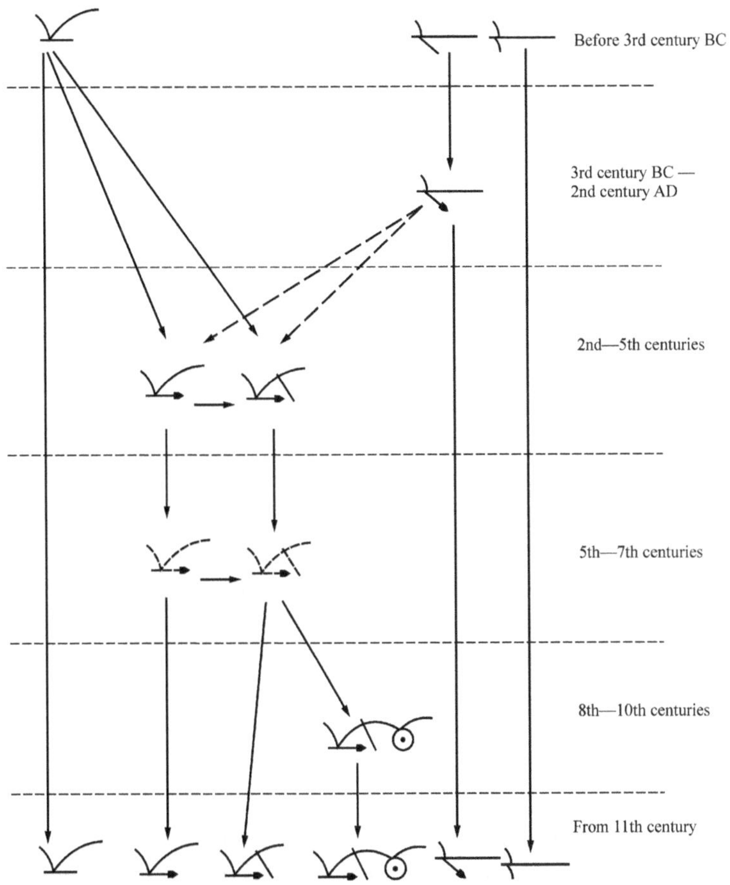

FIGURE 2.5
The development of arable tools. Tools dated to the period, for which plowshare finds are known, are marked with a dashed line (the third quarter of the first millennium AD)

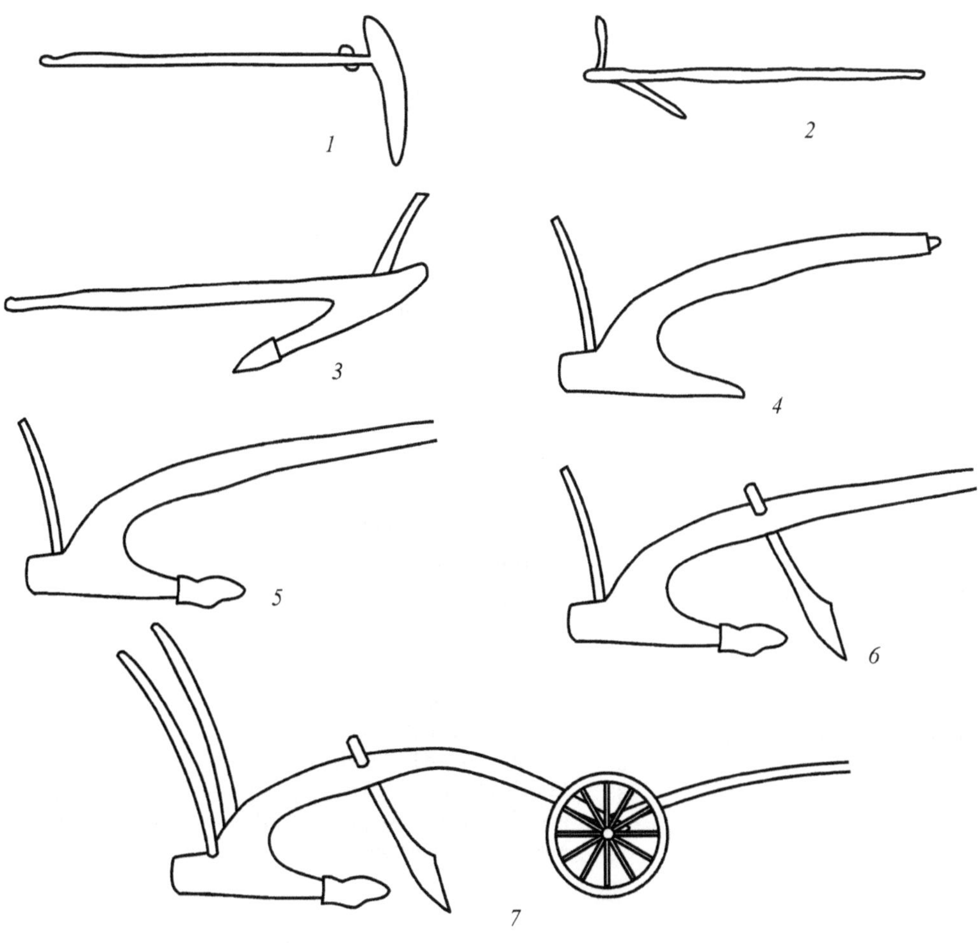

FIGURE 2.6
Possible varieties of arable tools of the first millennium (1–7)

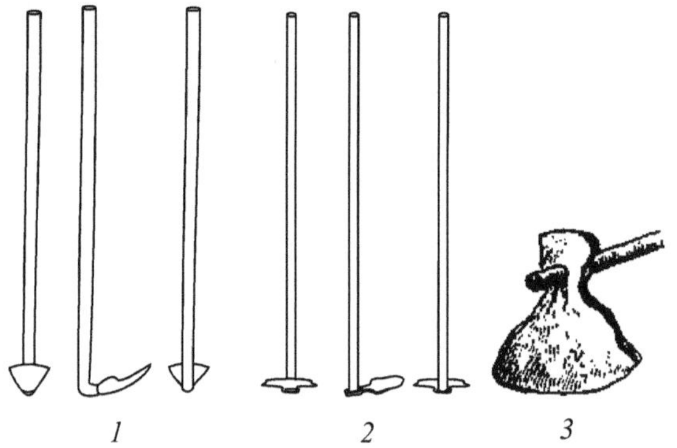

FIGURE 2.7
General forms of hoes: *1* — with vertical socket; *2* — with horizontal socket; *3* — attachment of the hoe with horizontal socket

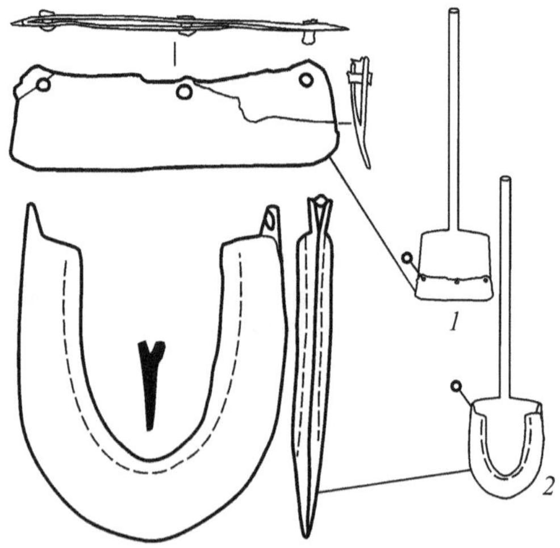

FIGURE 2.8
General forms of spades: *1* — with rectangular blade; *2* — with semicircular blade

new material is compared to similar samples that have already been identified, and on the basis of a wide range of publications on the subject. Imprints of weed seeds and of fruits or plant remains other than seeds have been identified on the basis of a catalogue of weeds.[28]

Because the search for, and selection of palaeobotanical samples from Saltiv sites have started just recently, and now they have become quite sporadic, the main palaeobotanical database for the Saltic culture consists of information obtained from pottery imprints.[29] Using plant material (including seeds) as temper is a relatively well-known technology of for pottery manufacture in communities with strong agricultural traditions. The role of the plant remains (as of any other temper) was to reduce shrinkage of the pugged clay. Sometimes, however, mixing of grains in the paste must have been symbolic. The introduction of such a temper also accelerated and improved drying and firing.[30]

For us, it is important that grains were added to the temper only sporadically, and without preference for any particular cereal (except proso millet, which is discussed below). The random character of the cereal selection for temper makes it possible to obtain statistically undistorted data. It should be also added, that according to Galina Pashkevich, the data obtained from flotation correlate well with data obtained from pottery imprints.[31]

Appropriate concepts are needed for the statistical analysis of the material. For example, the concept of "palaeobotanical complex" (PBC) designates the aggregate grains and seeds (imprints or charred material) characterizing a particular archaeological culture. It was possible to distinguish PBC on the basis of the archaeological material, i.e., material from closed finds with certain cultural attribution. A "palaeobotanical spectrum" results from the excavation of site with single layer (occupation phase). While PBC is associated with settlement features (dwellings, pits), PBS concerns the settlement site as a whole. However, the PBS of different sites of the same archaeological culture constitute the PBC of that culture. The PBC of any given culture may have local variants, leading to the introduction of the notion of the "zonal PBC" of that culture.[32] The significance of crops within the PBC is indicated by percentages. Of course, that percentage can only approximate the real share of crop in the household; however, at the current state of palaeobotanical studies, that remains the only standard by which one can compare the role of various groups among different archaeological cultures.[33]

While the material is presented in a quantitative / percentage ratio (PBS_q), a number of conditions are taken into account. In the case of the PBS of sites, where the only palaeobotanical data derives from pottery imprints, we removed imprints of proso millet grains on the bottom of pots from the analysis, in order to eliminate substantial statistical errors. Pot bottoms may have an unusually large number of imprints, because grains were spread on the working bench or table to prevent the wet clay from

28 Veselovskii, Lysenko, and Man'ko 1988.
29 Gorbanenko 2015b and 2016 P'iatnyts'ke.

30 Bobrinskii 1978, p. 101–102.
31 Kravchenko and Pashkevich 1985.
32 Kravchenko and Pashkevich 1985, 180.
33 Kravchenko and Pashkevich 1985. Lebedeva 2008, 97 has suggested an alternative phrase ("archaeobotanical spectrum," or ABS) for the same thing.

sticking to it.[34] Proso millet grains were most appropriate for the job, as they are the smallest cereal seeds.

The aim is to present the PBS of cultivated cereals as a measure of their possible cultivation (or consumption), but not in terms of the number of grains, the weight of which is significantly different from one crop to the other. For that reason, we have listed PBS by weight (PBS_m, see below). Scholars usually pay attention to the quantitative ratio, but neglect grain peculiarities for each crop. In fact, different cereals have quite different seeds.[35] Other scholars also have noted the importance of this observation.[36]

For an entire decade (2002–2012), we have upheld this purely hypothetical position. However, taking into account the difference in size between different cereal seeds does go a long way to meet the demand for a more nuanced interpretation of the results, seed size is not a good measure of comparison between crops. After all, in most written sources (both medieval and modern), crops are sold by weight, not by seed size.

On the other hand, the "archaeological" material is not really fit for a true comparison between crops. With potter imprints, one can only approximate the size of the seeds. Charred seeds are also not appropriate for comparisons based on size, as different crop seeds burn differently. That makes weighing charred seeds pointless. The only way out of the conundrum is to compare the grain amount for various crops obtained within one year, in samples that are of the same weight. For that, we weighed 10-gram samples of grains of the main crops (proso millet; hulled barley; hard wheat to characterize emmer wheat; soft wheat to characterize naked variations of cereals; rye; oats), and listed the contents of each sample. The following results were obtained:

proso millet	1285
hulled barley	232
hard wheat	207
soft wheat	227
rye	266
oats	375

Taking proso millet, which is the smallest grain, as a basic unit of measurement, we then counted grains to determine the index (one grain of certain cultivated plant is *n* of proso millet grain):

proso millet	1
hulled barley	5.5
hard wheat	6.2
soft wheat	5.7
rye	4.8
oats	3.4

PBS_m and PBC_m were calculated on that basis for further interpretations.[37]

Some crops included in the PBS and PBC offer little information for further analysis. By contrast, others provide valuable information for further interpretation, in addition to establishing the fact that they were cultivated in ancient times.

A large amount of proso millet (*Panicum milliaceum*) may indicate the seeding after fire (slash-and-burn agriculture), the use of floodplain areas, an overall low level of agricultural technology, with extensive forms of farming, or the use of rainfed farming systems (in which seeding is exposed to rain on a pre-plowed land). Ethnographic data confirm the correlation between swidden cultivation and the prevalence of proso millet, along with other resilient crops, such as barley and emmer wheat.[38] Millet, on the other hand, is characterized by low growth in its initial growing season. Weeding is absolutely necessary, therefore, for the survival of the seedlings. Deep plowing does not work for millet.[39] Swidden cultivation implies sowing immediately after the fire or after a slight loosening of the soil, with no need for deep plowing. Moreover, because of the delay in germination of proso millet at the beginning of the growing season, deep plowing could actually be detrimental to germination.

The presence of proso millet as a sign of rainfed farming system is most characteristic for Khazaria, especially because part of forest-steppe population of Khaganate practiced a form of semi-nomadism that combined transhumance and arable farming. Soil tillage cultivation could support both components of that economic mode, or both occupations were equally depending upon life circumstances.[40] Therefore, plots located at a considerable distance from each other, in river valleys or in the steppe, were cultivated sporadically, whenever the human group moved to that specific location. Historical data show such seasonal fields being cultivated by Tatars in the region of the Sea of Azov, particularly those along the Donets,

34 Bobrinskii 1978, 39.
35 Pashkevich and Gorbanenko 2002–2003, 161–162; Gorbanenko 2007, 24.
36 Viazov 2001.
37 Gorbanenko 2012a, 169.
38 Tret'iakov 1932, 13–15.
39 Hrihorovich 1933, 7 and 8; Ielagin 1955, 9; Lysov 1968, 8.
40 Mikheev 1985, 43.

as well as by Crimean Tatars along the Oril and Samara rivers.[41] The best crop for fields cultivated sporadically is proso millet. The population of the forest-steppe zone of Khazaria may have also used that crop in the floodplains. Given the relatively well-developed crafts in the Saltiv culture, it is unlikely that the presence of proso millet on sites in Khazaria indicates a low level of agricultural technology.

Similarly, a large amount of barley (*Hordeum vulgare*) within any given palaeobotanical sample may indicate slash and burn clearing (much like with proso millet and emmer wheat; see above),[42] a low level of farming technology, or a partial subordination of farming to the needs of animal husbandry. The correlation between barley and a relatively low level of agricultural technology may be confirmed by agronomic properties of the crop. After all, this cereal is resistant to variable climate conditions and soil fertility, drought-resistant, and precocious (with a growing season of 60 to 110 days).[43] In the regions to the south with a longer growing season, one can sometimes count on two harvests per year. In Georgia, barley can be sown on wheat fields, if that cereal did not germinate.[44] Barley could also be used as fodder, both for horses and for pigs. Indeed, the straw and chaff of barley have nutritional qualities that bring them very closer to hay.[45] Although the population of the forest-steppe region of Khazaria may have used slash-and-burn to increase the cultivated areas, it is most likely that the presence of barley on sites of the Saltiv culture is an indication that the crop was cultivated for fodder, and thus indirectly points to animal husbandry.

A large amount of hulled (emmer) wheat (*Triticum dicoccon*) indicates either resistance to innovation and the persistence of the traditional forms of agriculture, or the use of slash-and-burn (along with proso millet and barley, see above).[46] Together with barley, emmer wheat was one of the first domesticated crops, which appeared on the territory of modern Ukraine among agricultural populations. Since its appearance and until the end of the first millennium AD, this crop was one of the main cultivated cereals,[47] best adapted to climate and soil conditions.[48]

A large amount of wheat (*Triticum aestivum* s. l.) indicates either a high level of agricultural technology or agricultural intensification (the pursuit of higher yields). The first option may become apparent when one compares the time of increased importance of naked forms of wheat with the overall development of farming during the last quarter of the first millennium.[49] The second option is substantiated by the information from written sources pertaining to the historical development of the Slavs (the rise of Kievan Rus') and the population of the Khazar Khaganate. Both options are backed by the agrobiological properties of the crop. Naked wheat requires soil treatment (deep plowing) in order to produced yields higher than any of the other cereals.[50]

A large amount of rye (*Secale cereale*) in the palaeobotanical sample suggests either a high level of agricultural technology (along with naked wheat) or a partial subordination of farming to the needs of animal husbandry (along with a significant share of barley and/or oats).

The first option has at one point preoccupied scholars interested in ancient farming.[51] Much like naked wheat, rye can give high yields, but it is more capricious, and dependent upon environmental conditions and the quality of the soil.[52] But rye could definitely be used as fodder over the winter.

Common oats (*Avena sativa*) poses problem of interpretation. Oats appear only occasionally and in small percentage in palaeobotanical samples dated between the late first millennium BC and the late first millennium AD,[53] or even later.[54]

Oats grain finds may be evidence of the partial subordination of farming to the needs of animal husbandry (along with barley and rye). Oats may indeed be used as fodder. At Rohalyk, most grains are of hulled and hulled "bottle-shaped" barley (*Hordeum vulgare* var. *laguncuIiforme*), as well as oats.[55] This has rightly been interpreted as a sign of the latter crop being cultivated for fodder, given the important role of animal husbandry in the life of Saltiv communities.[56] We drew a similar conclusion from the analysis of palaeobotanical samples from Saltiv site in the Donets region, where oats appear in relatively low percentages (although slightly higher than other

41 Herberstein 1866, 153; Skrzhinskaia 1971, 150.
42 Tret'iakov 1932, 13–15.
43 Kobylianskii and Luk'ianova 1990, 188.
44 Bregadze 1982, p. 81.
45 Vavilov 1986, 124.
46 Tret'iakov 1932, 13–15.
47 Ianushevich 1976 and 1986; Pashkevich 1991a, 1991b, 1991c, 1991d, 1992, 1993; Pashkevich and Videiko 2006.
48 Brezhnev 1979, 50–52, 213, and 214.

49 Gorbanenko and Pashkevich 2010, 103–276.
50 Brezhnev 1979, 50–52, 213, and 214.
51 Lange 1975; Iazhdzhevskii 1988.
52 Degtiareva 1981, 22–25; Kobylianskii1989, 276.
53 Pashkevich and Gorbanenko 2010.
54 Pashkevich 1991d.
55 Pashkevich, Gorbanenko 2002c; 2004.
56 Mikheiev 1985, p. 26.

cereals) together with barley and/or rye in high percentages. This cannot be interpreted as an indication of a low level of farming, and must indicate the cultivation of oats for fodder.[57]

The presence of weeds in palaeobotanic samples is generally interpreted as an indication of old arable fields. Ethnographic observations confirm that weeds do not typically appear in places cleared by fire (slash-and-burn kills all vegetation, not just trees).[58] Weeds are also not present in floodplains, because of floods. Virgin parcels of land were commonly free of weeds. Cockspur grass (also known as barnyard millet, *Echinochloa cruss-galli*) and wild millet (also known as yellow foxtail, *Setaria pumila*) are two summer annual grasses commonly found in fields cultivated with mille, where they typically exhaust the soil. They are also found with other, as well as near dwellings and along roads. One of the most detrimental weeds is the black-bindweed (*Fallopia convolvulus*). Twining about the stems of cultivated plants, the weed forces them to the ground, thus making harvesting very difficult. Winter weeds such as rye (*Bromus secalinus*) and field brome (*Bromus arvensis*) usually out-compete winter rye and wheat. Lamb's quarters (*Chenopodium album*), and field bindweed (*Convolvulus arvensis*) are with crops of all cereals, but also next to dwellings, in garbage dumps, along roads, and on pastures. Annual such as cleavers (*Galium aparine*), which occur in spring and winter crop sowings and in vegetable gardens, are particularly harmful, as one single plant produces up to 1,000 fruits.[59]

Of all weeds, brome (*Bromus* sp.) — both rye and field brome — offers the most important information for interpretation. Its presence in the Middle Ages can serve as proof of the cultivation of rye, of the use of old arable lands, of the planting of both spring and winter crops, and of the two- or/and three-field system. Most scholars agree that rye and field brome signal the cultivation of winter rye.[60] As brome is a weed of the *winter* rye, its presence is simultaneously an indication of two- or three-field rotation. During the last quarter of the first millennium, the Slavs, according to Ibrahim ibn Yaqub, "sow in two seasons, in the summer and the spring, and reap two harvests."[61] This may also apply to the neighbors of the Eastern Slavs, the inhabitants of the forest-steppe region of Khazaria, especially because they were actually more advanced than the Slavs in matters of agriculture.[62] They planted spring cereals in spring and winter cereals in late summer or autumn, a practice that is confirmed by 19th-century ethnographic reports. According to those reports, with the three-field system, the first part of the field was plowed in spring for spring crops; the second part was plowed in summer for lying fallow; plowing for the winter crops was carried out in early autumn.[63]

Combining the results of the analysis of the potential resource zone of any given settlement, the tools for soil tillage and the PBS, it becomes possible to characterize the overall development of farming, as well as the use of different plots of land for various agricultural needs.[64]

When analyzing the productivity of harvesting tools (sickles, *gorbushy*, and scythes), scholars often take into account the following parameters of any sickle: the total length of the tool, the length of blade and the arc, the length of the handle (for socketed specimens), the top and height of the blade arc.[65] Those indicators are regarded as sufficient to gauge the technological progress of harvesting tools (sickles), but differences in the development of agricultural tools among various archaeological cultures are also assessed on the basis of how the handle is attached to the blade — hooked, socketed, shanked, stemmed, or folding (toggle-action) (Fig. 2.9).[66] Without any doubt, sickle finds show that cereals were grown (and not just used) locally. The development of agriculture is therefore indirectly tied to the parameters of the tools, but also to the total number of finds.

Gorbushy scythes could be used both for harvesting cereals with tight spikes and for haymaking in order to feed livestock during the winter. Both archaeological and ethnographic studies have made it possible to gain an overall picture of how ancient scythes were used and changed (Fig. 2.10).[67] For that reconstruction it is important also to consider how the blade was attached to the handle (Fig. 2.11). Scythes only demonstrate an agricultural focus

57 Koloda and Gorbanenko 2010.
58 Tret'iakov 1932, 4–6.
59 See Vasiulina 1970; Veselovskii, Lysenko, and Man'ko 1988.
60 Smirnov and Sosnikhina 1984, 5–7. See also Kir'ianov 1959, 333; Kir'ianov 1967, 177; Mykhaylyna, Pashkevich, and Pyvovarov 2007, 60.
61 Izvestiia 1878, p. 54.
62 For more details, see Chapter 6.
63 Parfenov 1873, 642–643.
64 For examples of such characterizations for Saltiv sites, see Koloda and Gorbanenko 2009; Koloda and Gorbanenko 2010; Gorbanenko, Koloda, and Pashkevich 2009; Gorbanenko and Koloda 2010.
65 Levashova 1956, 60–95. *Gorbusha* (*gorbushy*, in the plural) is a typically East European scythe with a short and slightly curved handle.
66 Minasian 1978b.
67 Archaeology: Miroliubov 1976, 122 and 123; Mikheev 1985, 29–31. Ethnography: Guldenstedt 1804, 37 and 38.

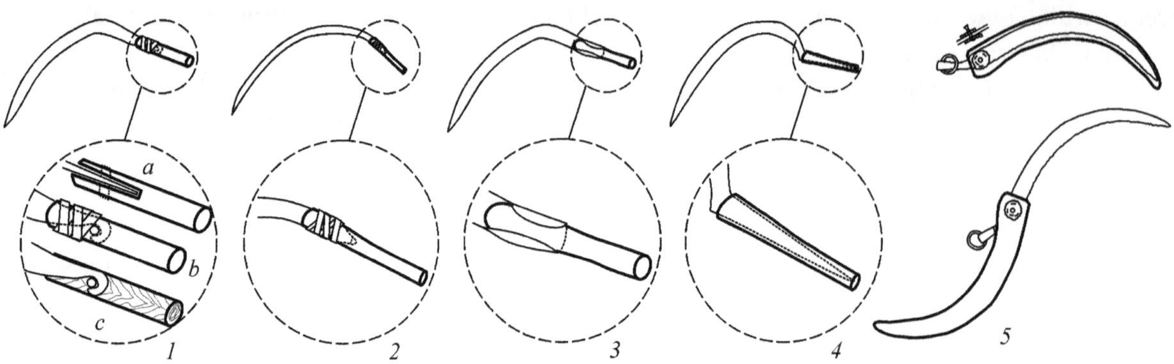

FIGURE 2.9 Types of attachment of sickle handles: 1 — hooked; 2 — shanked; 3 — socketed; 4 — stemmed; 5 — folding

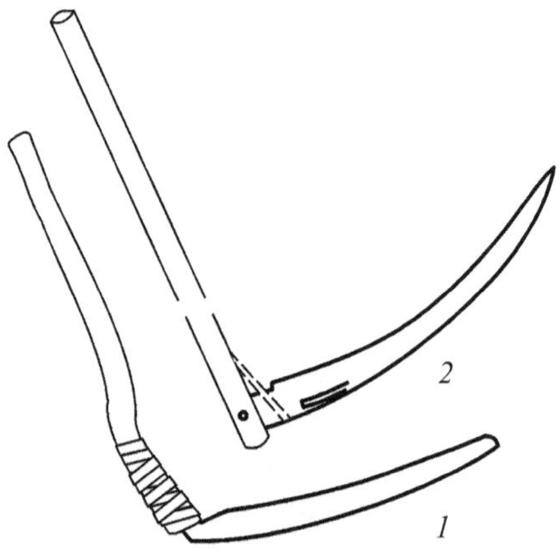

FIGURE 2.10
General forms of scythes: 1 — type I — *gorbusha* scythe; 2 — type II (according to the classification of V. K. Mikheev)

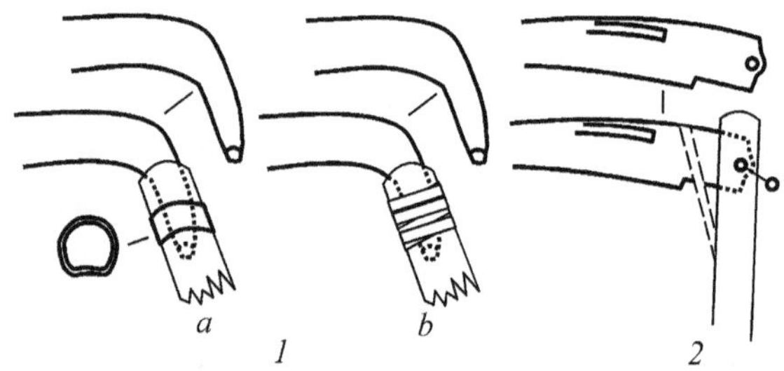

FIGURE 2.11
Modes of attaching the scythe blade to the snath: 1 — with heel: *a* — with ring; *b* — with a flexible strap; 2 — without heel with a hole

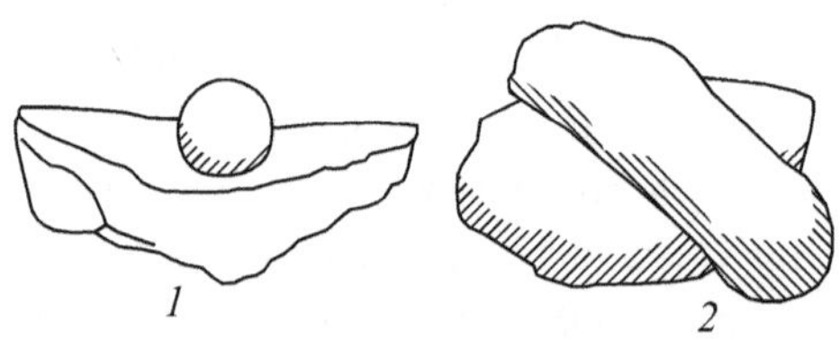

FIGURE 2.12
Types of graters: 1 — one-handed; 2 — two-handed

of economy broadly defined, with no particulars being assessed.

Tools for processing food (grains) may be classified based on the mode of operation. For example, there are one- and two-handed graters, depending upon the use of one hand or of both (Fig. 2.12).[68] Such tools were used for grinding seeds, but also for processing grains for groats and flour. Where millstones were available, graters were probably of secondary significance. Rauf Minasian's studies and the ethnographic reports make it possible to reconstruct the manual, rotary millstones of the early Middle Ages (Fig. 2.13).[69] Tools for food processing indicate cereal consumption, but cannot in any way be associated with soil cultivation. Nevertheless, the idea that a widespread use of millstones was associated with a considerable improvement of farming and high higher yields appears to be true.

1.2 *Animal Husbandry*

A necessary premise for determining the place of animal husbandry in ancient economy is a sufficient amount of material for analysis. At least 500 bones must be in the sample before any analysis is attempted, as any determination of species leads to statistical considerations.[70] At the same time, that same minimum may probably be advanced for any attempt to establish a statistically sufficient number of animals.

Unlike palaeobotany, zooarchaeology has a long history of research, reaching back to the mid-20th century. Over time, several methods were used. Initially, scholars used mainly methods developed in the former USSR, but

68 Ponomarev 1955, 17–19; Krasnov 1971, 83.
69 Minasian 1978a.

70 Zhuravlev 1991.

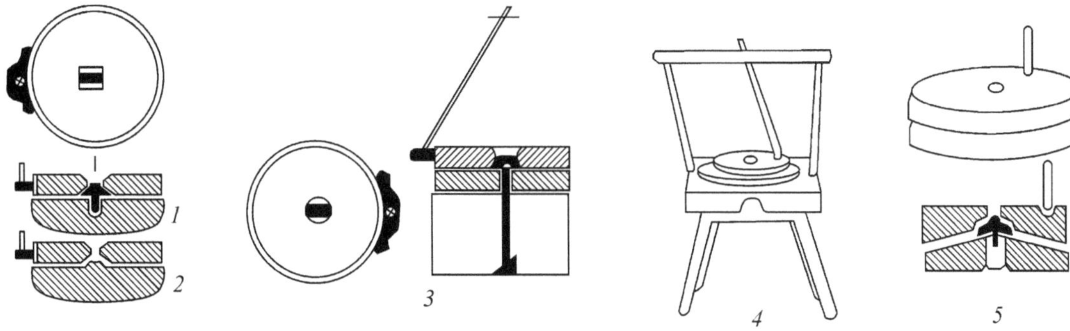

FIGURE 2.13 Reconstruction of millstones, after R. S. Minasian: *1* — group II, variety A; *2* — group II, variety B; *3* — group I; *4* — group III, variety A; *5* — group III, variety B

new methodological studies have allowed for an increase precision of the identifications.[71]

The zooarchaeological corpus pertaining to animal husbandry in the forest-steppe region of Khazaria includes over 2,000 bone fragments. All of them have been identified, despite the considerable fragmentation (some of the bones had imprints of plant roots). The identification of bone remains of small livestock was based on the morphological criteria advanced by various authors.[72] For measuring the bones we relied on the method developed by the Institute for Paleoanatomy, Domestication Research and History of Veterinary Medicine of the University of Munich.[73] A minimal number of specimens per taxon (NISP) was calculated, first of all, on the basis of isolated teeth and jaw fragments. In addition to right and left sides, other conditions of ontogenetic development of the dental system were taken into account, namely the degree of teething, growth and formation of one tooth or another, as well as degree of their deterioration. In some cases, the number of animals was based on the most complete of the preserved limb bones, more often than not talus bones. The method has already been implemented in several studies pertaining to faunal assemblages from early medieval sites of the contact zone between the East Slavs (the Severians of the Russian Primary Chronicle) and the population of the northwestern region of Khazaria.[74]

From the beginning of zooarchaeology as an academic discipline, most scholars employed the minimum number of animals (MNA), calculated for every site. However, there are clear problems with this number and the associated method.[75] Others, rejecting MNA, employ percentages of animal bones, thus equating those indicators to the number of animals. The alternative is equally open to criticism.[76] We have decided to rely on MNA for the zooarchaeological component of this book.

One of the important interpretive steps is calculated the weight of every animal. Such concerns go back to the beginnings of zooarchaeological studies.[77] The best foundation for such studies is the calculation based on materials deriving directly from a given site. However, such data are not always available, as not all (zoo)archaeologists carry out such calculations. We have therefore decided to use instead average indicators from several sites, where the starting unit was taken from the smallest domestic animals, namely small livestock with an average weight of 40 kg.[78] On that basis, conversion factors of weight of other animals have been proposed: small livestock — 1; cattle — 6; horse — 5.5; pig — 1.5.[79] For similar reasons, namely the incomplete character of the data, one should also abandon the idea of weight gradations based on age. Behind this concern with weight is the desire to estimate as accurately as possible the amount of meat available at any one time from domesticated animals. Since the data are not always sufficient for accurate estimates, one needs to rely on relative data (as in the case of the PBS_m), presented as percentages.

71 For Soviet-era zooarchaeology, see Gromova 1950, 1953, and 1960.

72 Gromova 1950; Fernandez 2001; Halstead, Collins, and Isaakidou 2002.

73 Driesch 1976.

74 Kroitor and Koloda 2015 and 2016.

75 Antipina 1997 and 2008.

76 Taphonomic considerations complicate the picture considerably. Most scholars, however, simply assume that on settlement sites, the most common bones are those of cattle.

77 Pidoplichko 1937; Timchenko 1972; Zhuravlev 2001; Antipina 2007 and 2008.

78 Antipina 2007, 299.

79 Antipina 2008, 71.

Grazing could take place in the pasture either locally or through transhumance. A grazing system based on pastures implies that animals are returned to the settlement every day to spend the night there. In a transhumant system, animals are driven away in the warm season to distant pastures, and they return to the settlement in winter. The analysis of potential resource areas near any given settlement can lead to identification of the particular system in use at any given time. It is of course quite possible that both variants were used simultaneously, holding a certain number of certain animals nearby, in order to replenish stocks of fresh milk.

The study of animal husbandry thus leads to a number of possible conclusions. Depending upon environmental conditions, one can identify areas for grazing and thus the grazing system in use. The age ratio for different animals in the herd may indicate the feeding base capacity (a smaller number of young animals points to better feeding base). The ratio of animals in the herd, in general, makes it possible to assess the significance of animal husbandry. Using Tsalkin's observation, one can thus evaluate productive forces for farming.

The ratio of species in a herd and the varieties of grain in PBS can give a good indication of whether and how one branch of agriculture was subordinated to the other. In addition to zooarchaeological data, the study of animal husbandry also relies on tool finds, either those directly linked to animal breeding (cattle bells or shears) or those involved in slaughtering (butchering knives). To be sure, such data rarely play any significant role in characterizing animal husbandry, as they are often used to illustrate conclusions that have already been drawn on the basis of the bone material.

Any attempt to evaluate the development of ancient agriculture needs therefore to take into consideration all of the materials available, even when fragmentary. The relative analytical power of various methods of research remains to be demonstrated, but only the future development of the various branches of scientific knowledge, and the cooperation of various specialists could increase the potential of the model of interpretation advance in this book to enrich our knowledge of individual sites and their agricultural profile.

CHAPTER 3

Natural Conditions and Resettlement Areas

The economy of ancient societies, especially agriculture in its traditional forms, was largely dependent on natural and climate conditions and on features of the natural landscape around settlements. The latter served as a base for organization of agricultural production in a certain area, a principal place of production of appropriate working tools, as well as a center for storing and processing of the main products of agricultural activities.

1 Natural Conditions

The chronological correction of natural and climate conditions during the historical period.[1]

Natural conditions were until very recently the most important factor in agricultural production, responsible not only for the direction of development, but also for the crops chose for cultivation and the breeds selected for animal husbandry. It is therefore most appropriate to being the study of agriculture in the forest-steppe region between the Dnieper and the Don at the end of the first millennium AD, we with natural conditions and the climate, as well as their interrelations.

It is crucial to match archaeological chronologies (especially the period of the Saltiv culture) with the time scale of climate (Fig. 3.1). The Saltiv culture, which flourished during the last quarter of the first millennium (from the mid-8th to the mid-10th century) coincides with the late Holocene. Scholars have divided the Holocene into short climate phases. Several climate periodizations have been advanced for the Holocene. In Ukraine, the most prominent is that of N. A. Khotinskii,[2] combined to M. F. Veklych's studies.[3] This combination is the basis for more recent periodizations of the Holocene.[4]

M. F. Veklych operates with the following conventions: *hl* — Holocene; *c* — nanostage of the first subordination (period); a first Arabic numeral — nanostage of the second subordination (sub-period); a second Arabic numeral — (after the dash) — micro-stage; odd numbers — the time of global cooling, even numbers — warming.[5] Veklych's scheme is global, but it can be used to create regional and local schemes of environmental development in the Holocene, since the planetary stages are manifested in every region of the Earth, albeit with local peculiarities.[6]

Others, however, have advanced different points of view on the global synchronization of natural climate events in the Holocene. For example, Aleksandr Aleksandrovskii and Elena Aleksandrovskaia propose that moisture was dominant in the Atlantic period of the Holocene, when some regions in Eurasia had a wet climate, while in others the climate was drier than today.[7] A number of studies have meanwhile confirmed the diversity of the climatic changes in different regions of the Earth during the Middle and the Late Holocene.[8]

In Khotinskii's chart, the Holocene Subatlantic (SA) covers the last 2,800 years of environmental history. The historical interval that we consider in this book coincides with the second phase of the Holocene Subatlantic (SA-2), 1,600 to 800 years ago. Khotinskii distinguishes two stages — SA-2A (1,600 to 1,200 years ago) and SA-2B (1,200 to 800 years ago). The Saltiv archaeological culture flourished during stage SA-2B (Fig. 3.1).

Of great interest is the reconstruction of natural conditions in the central region of Eastern Europe, with in the forest-steppe belt, where economic activity of the Saltiv communities took place. Some general idea about the natural environment may be obtained from the analysis of literary sources. On the other hand, the chronological interval 1,200 to 800 years ago is that of the Medieval Warm Period.[9] In the Central Russian Upland, The Medieval Warm Period brought a stabilization of the geomorphological systems, without slope erosion and decrease in the intensity of alluvial accumulation. Between 1,440 and 1,060 years ago (dates of buried soils), a process of soil formation continuously took place on the surface of floodplains and on slopes in river valleys, as well as on slopes and on bottoms of ravines. This process is responsible

1 We would like to thank Iurii G. Chendev for his advice and clarifications on palaeogeography.
2 Khotinskii 1977.
3 Khotinskii 1977; Veklych 1987.
4 Matviishyna 2010, 132–133.
5 Those conventions vary from one publication to the other (Veklych 1987; Veklych and Gerasimenkov 1993; Matviishyna 2010). For this book, however, we have adopted the conventions in Veklych 1987.
6 Veklych and Gerasimenko 1993, 86–87.
7 Aleksandrovskii and Aleksandrovskaia 2005, 182.
8 Chendev 2008, 12–16.
9 Borisenkov and Pasetskii 1983.

 | DOI:10.1163/9789004429574_005

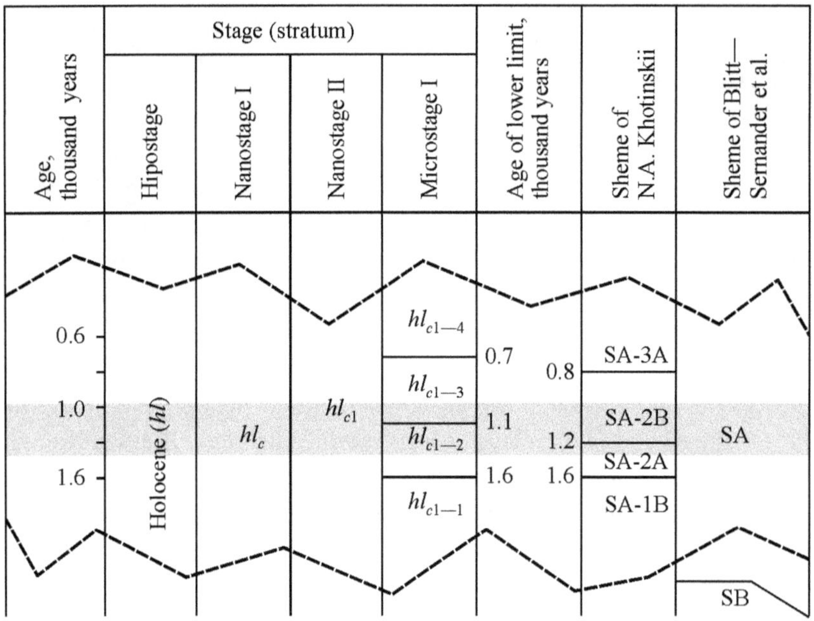

FIGURE 3.1
Holocene phases (based on the fragment of chronological scheme from: Matviishyna 2010, 132, plate 4.1). The period of the Saltiv culture is shown in gray (made by S. A. Gorbanenko)

for one of five Holocene soils identified in depressions of natural erosion.[10]

1.1 *Natural Zone*

The area under study corresponds is located in the forest-steppe zone at the center of Eastern Europe. The sites are located in the southern part of the forest-steppe zone, at different distances (from 5–10 to 40–50 km) from the northern boundary of the steppe zone, which runs from west to east from the source of the Orchik to the sources of the Berestova and Bereka rivers. The boundary crosses the Donets near the present-day town of Balakliia, then turning northeast, passes through the upper reaches of Velyky Burluk, where it runs approximately between Valuiki and Liski.[11] In terms of landscape zones, the western part of this area is within the Left-Bank Dnieper (Livoberezhnodniprovska) forest-steppe province, while the eastern one belongs to the central Russian forest-steppe province. The Left-Bank Dnieper province comprises two elevated physical-geographic areas — the northern and southern Poltava (Pivnichnopoltavska and Pivdennopoltavska) areas. The steppe zone is located outside the area of the sites under study. At the southern border of the forest-steppe belt natural landscapes of northern steppe belt appear: the Left-Bank Dnieper-Sea of Azov (Livoberezhnodniprovsko-Priazovska) northern steppe province, the Donets (Donetska) northern steppe province, the Western Donets (Zakhidnodonetska) highlands, and the Donets-Don (Zadonetsko-Donska) northern steppe province. The forest-steppe landscapes in the territory under study territory are typically deciduous forests of the Kharkiv type and steppe meadows of the Kolomak type.[12]

1.2 *Relief*

The forest-steppe region is a land of rolling hills.[13] The series of low hills was created by valleys and ravines cutting through the plain area. This is particularly true for the watershed plateau, which is cut by ravines branching in many directions, and covered by forest.[14] Three types of relief predominate: watershed plateaus, gently sloping; sloping and steep banks of rivers and ravines; and lowlands near rivers. Dissected loess plains are common in the northern part of central Russian forest-steppe province. In the south, the landscape is dominated by highly dissected, loess relict hilly elevations and their slopes with deep gullies, ravines and landslides. Two denudation levels are predominant: Neogene, which is about 200 m high, and Paleogene, with an altitude of less than 200 m. River valleys with downcutting depth of 100 to 150 m are quite typical. A network of gullies and ravines is well branched (dissection density is 1.1 to 1.5 km/sq km). A combination of young (gullies, ravines, and river valleys) and old (channels of glacier runoff) forms is characteristic. Tectonic movements and erosion played a significant role in the formation of the current relief.[15]

10 Sycheva and Chichagova 1999.
11 Milkov 1950, 179.
12 Gerenchuk 1964, 9.
13 Popov, Marynych, and Lan'ko 1968, 226.
14 Monin 1957, 172.
15 Babychev and Marynych 1993, 178.

The highlands of Kharkiv are indicated by a general inclination of the surface to the south and southeast, by the spread of loessial rocks, and by the dissection of relief to a depth of 50 to 100 m along the valleys of the Donets river and of its tributaries (the Mzha and the Udi rivers). The landscape is dominated by rolling watershed hills with black earth; valleys and ravines with floodplains, forests on the bottom and slopes of ravines, and sod-gley and gray forest soils; and ravines with black earth and gray forest soils. In addition, there are floodplain, floodplain -terraced, and sand-marsh spots in the valleys of rivers which have alluvial, meadow, meadow-black earth, and humus surface-gleyed soils.[16]

The relief described above has a considerable influence upon the local climate, vegetation, soils, and soil fertility. Their various combinations in the region under study led to components of the environment that are different from one archaeological site to the other.

1.3 *The Climate*

Of the territory under study is moderately continental. Winters are relatively mild, with average January temperature of 7.5 centigrades, frequent thaws, and unstable snow cover. Summers are moderately warm with sufficient rainfall during the growing season, and July temperatures between 18 and 21 centigrades. Average annual temperatures are between 6.5 and 7.5 centigrades, while average precipitation levels run from 500 to 700 mm annually.[17]

It has long been noted that despite the relatively high level of farming technology, that was not sufficient to offset "the adverse effects that natural phenomena had on yields."[18] This is important to keep in mind even if, as scholars seem to agree, at the end of the first millennium, the climate in Eastern Europe was not very different from the current situation.[19] However, according to recent data, there were variations in the climatic conditions of both periods, which could in principle have influenced different sectors of human activity. The climate of the Holocene changed several times. Warming periods alternated with periods of cooling, much like wet and dry regimes. During the last third of the Holocene, about 4,500 years ago (the Subboreal and the Subatlantic phases), climate in Eurasia became cooler and wetter, with slight variations towards warming. Climatic fluctuations were typical for the Subatlantic as according to palaeogeographers they were associated with the cycles of solar activity. On the basis of the quantity of C-14 in the atmosphere, the cyclic recurrence happens every 200 years or so.[20]

Between 3,200 to 1,000 years ago (after the climatic optimum of the Atlantic, the wet and relatively wet phase of the Holocene), because of the degradation of deciduous trees and the developments of dark coniferous (spruce) forests, the northwestern part of the Russian Plain experienced increased humidity and subsequent cooling.[21] In the forest-steppe belt, at the center of Eastern Europe, the wet phase of the Late Holocene was accompanied by the growth of the forest into the steppe land and thus by the extension of the forest-steppe into the steppe belt.[22]

One of the significant cooling phases of the first millennium AD falls approximately in its middle, between SA-2A and SA-2B. The latter coincided with a gradual warming in almost all of Eastern Europe, as well as in the arctic, subarctic, and temperate zones of the Northern Hemisphere, in general.[23] This reconstruction of the temperature changes is confirmed by data from other parts of the Northern Hemisphere.[24] However, palaeogeographic reconstructions suggest that the Saltiv culture coincided with a climate change that manifested itself in a lower level of annual precipitation (reduced by 50 to 70 mm between the mid-8th and the mid-10th century), and slightly annual average temperature (an increase of no more than half of a centigrade; Fig. 3.2).

The 10th and 11th centuries marked the Thermal Maximum in Europe. At that time, communities in Kievan Rus' had to cope with frequent droughts.[25] Judging by the palaeogeographic reconstructions available, it seems therefore that the climate in the area under study was wetter at the beginning of the chronological span considered in this book than at its end.[26]

A reduction of the forest cover and an increase in the grass area indirectly confirm those changes. The increased aridity must have led to a reduction of the water table, immediately visible in both sources and wells. New studies have confirmed this reconstruction, particularly for the Donets region.[27] One needs to conclude, therefore, that at the time of the Saltiv culture, in the center of the Eastern

16 Babychev and Marynych 1993, 352.
17 Barbarych 1977, 18–49; Logvinov and Shcherban' 1984, 52 and 85.
18 Dovzhenok 1961, 183.
19 Borisov 1975, 77, 78, and 170.

20 Suess 1978.
21 Khotinskii 1977, 61, 163.
22 Aleksandrovskii and Aleksandrovskaia 2005, 181 and 198; Chendev 2008, 122.
23 Sleptsov and Klimenko 2005.
24 Burroughs 2005, 294 fig. 8.2.
25 Borisenkov and Pasetskii 1983.
26 Bezusko and Klymanov 1987.
27 Svistun and Chendev 2002–2003; Chendev and Koloda 2012; Koloda, Chendev, Bobrukova, and Dudin 2014. Similar results have been obtained for the neighboring regions of Psel-Vorskla

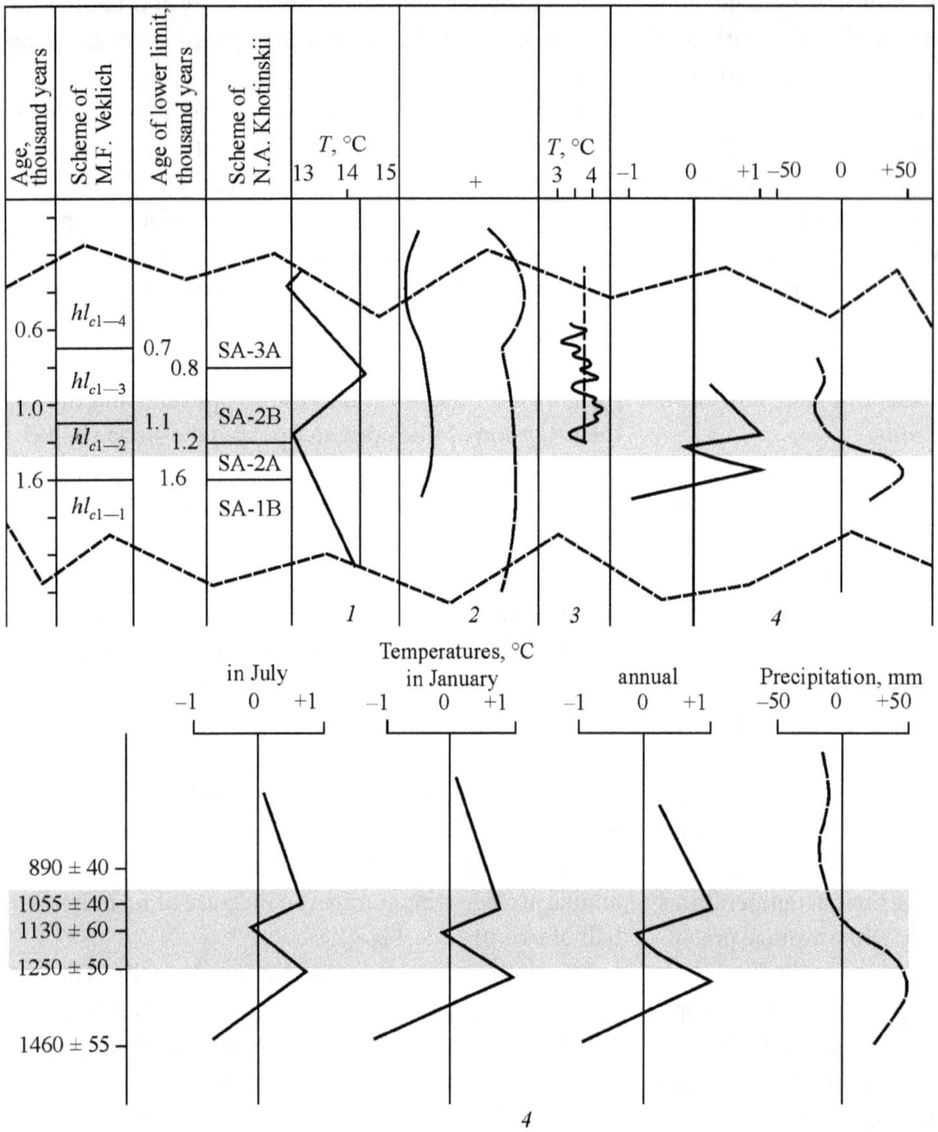

FIGURE 3.2
Climate change at the time of Saltiv culture (the period is shown in gray): *1* — global as compared to modern time (after Veklich 1987); *2* — the Russian Plain (after Khotinskii 1977); *3* — the average annual temperature in the center of the Russian Plain (after Klimenko and Sleptsov 1999, fig. 3); *4* — summarized paleoclimatic curves for the plains of western Ukraine (after Bezusko and Klymanov 1987). The solid line shows the temperature changes, the dashed one — amount of precipitation (made by S. A. Gorbanenko)

European forest-steppe zone, the climate conditions were quite stable (without obvious disasters), which have a favorable influence on economy.

1.4 *Water Content*

Characteristic for the Donets area, as well as for the forest-steppe region between the Dnieper and the Don, in general, is the ramification of the river system. Large rivers have large valleys and plain flow pattern. Small rivers (only a few dozens of kilometers long) often dry out in the summer.

In the lowlands, the watercourse of most river with slow flow changes periodically within the limits of the valley. Riverbeds changed little during the 20th and in the early 21st century, but in the historical past, the situation may have been different. In the lowlands, rivers changed the position of the watercourse over a relatively short period. Within 25 to 100 years, a riverbed may move across the entire width of the valley.[28] Pictures taken from the outer space reveal a large number of meanders, twists, old and new oxbows accompanying the current channel of the Donets. Particularly illustrative in this respect is the situation near the Mokhnach hillfort. The modern riverbed is about 1 km to east from the hillfort. However, according to the Plans of the General demarcation of 1785, at that time, the main channel of the river was on the western edge of the floodplain close to the plateau, a situation now observed near Koropovi Khutory.

Scholars agree that the coefficient of river network density for the Donets basin (0.13 km/sq. km) may be regarded as sufficient for the long-term use of that area for

(Chendev 2008) and Potudan-Tikhaia Sosna (Chendev and Sarapulkin 2012).

28 Boldakov 1951, 9–18.

the purpose of intensive economic activity.[29] However, one should take into consideration that in the recent past the anthropogenic factor has caused the degradation of river systems in the region. For example, over the last 200 years, the density of the river network in the Belgorod region of Russia decreased from 0.22 to 0.16 km/sq. km.[30] The river Mokhnachka, after which the nearby village was named, along with the hillfort located on its territory, appears Plan of the General demarcation of 1785. It became a creek and then disappeared in the late 20th century, as a result of the complete destruction of its channel through agricultural works.

It is reasonable to presume therefore that during the pre-industrial period the density of the river network for the Donets basin was between 0.17 and 0.18 km/sq. km. Because of the significant dissection of the terrain and relatively wet climate, a number of marshes were located in the landscape. The wetlands of the Donets basin appear only in the floodplains, where they account for 1 to 3 percent of the entire basin.[31]

For comparison it is worth mentioning that the river network density coefficient for the lands north of the Desna river is 0.29 km/sq. km, while in the south, in the more humid parts of present-day northern Ukraine is only 0.15 km/sq km.[32] In the upper and middle reaches of the Desna, waterlogging covers 20 percent of the entire area, but only 10 to 15 percent in the lower reaches. For the forest-steppe region of modern Ukraine the coefficient is no more than 5 to 10 percent. In the steppe zone, and partly in the lower reaches of the Vorskla and Psel rivers, the coefficient is 1 to 3 percent.[33]

In short, during the early Middle Ages, there was sufficient water in the region to sustain an intensive form of agriculture without irrigation.

1.5 *Vegetation*

Before the economic development that the region experienced during the last 250 to 300 years, this was a land of steppe meadows and deciduous forest areas distributed in a mosaic vegetation types throughout the vast forest-steppe zone. The trees most frequently growing in the forest are deciduous species, such as maple, linden, and oak in different combinations, which are typical for areas between river valleys.[34] In the Kharkiv geobotanical district (which is part of the Central Russian forest-steppe geobotanical province) about 40 percent of the total area was covered in the past by meadow steppes, which are now plowed. Interfluvial forest formations in that district are of the maple-linden-oak, linden-oak and oak types. On river terraces in the uplands, mixed oak-pine forests dominate, while in floodplains oak, groves and black poplar groves are common. A typical oak grove in the Kharkiv geobotanical district is the Homilsha National Park near Zmiiv.[35]

Various authors ascribe different flora variations to the southern boundary of forest-steppe region between the Dnieper and the Don. Some blame the (current) variations on the anthropogenic factor, particularly the intensive agricultural development of the region, others explain them in terms of natural factors. To be sure, until recently, both deciduous and mixed forests extended much farther to the south than the current boundary between the forest-steppe and the steppe belts. A large number of ravines as well as small river valleys (especially in the upper reaches) are still covered with relatively high (up to 30 m) deciduous forests with thick undergrowth. In the early 1990s, the landscape in the valley of the Bila river near the current administrative border between the Donetsk and the Luhans'k regions (next to the villages of Kruglyk, Gorodyshche, and Fashchivka) was dominated by huge oaks. According to some scholars, in the past forested areas along river valleys or adjacent ravines extended across the steppe all the way to shores of the Black Sea and the Sea of Azov.[36]

The study of hillforts in the basin of the river Donets, which show traces of occupation attributed to the Saltiv culture has revealed that the area was covered by forest in the late first millennium. The Mokhnach hillfort, for example, was surrounded by forests that were more likely larger than those of the Scythian period. In fact, the first (prehistoric) occupation phase of the hillfort coincided with heavy deforestation.[37] It is worth mentioning that during flotation fragments of hazelnut shells have been found (see Chapter 4), a good indication undergrowth in a deciduous forest. A very similar situation is documented for the Koropovi Khutory hillfort. During the Scythian period, the forest cover was much smaller than during the period of the Saltiv culture.[38]

The studies of samples of ancient soil buried underneath the medieval rampart at Verkhnyi Saltiv suggest that by the time the hillfort was built, the highland part of

29 Shvets, Drozd, and Levchenko 1957, 8.
30 Chendev and Petin 2006.
31 Shvets' 1946, 14 and 15.
32 Shvets', Drozd, and Levchenko 1957, 8.
33 Shvets' 1946, 14 and 15.
34 Babychev and Marynych 1993, 177.
35 Babychev and Marynych 1993, 357.
36 Dokuchaev 1949, 104.
37 Svistun and Chendev 2002–2003, 130.
38 Chendev and Koloda 2012.

the Donets valley was covered by deciduous forest.[39] The study of paleosols at P'iatnyts'ke I has equally indicated the presence of a deciduous forest immediately next to the settlement, and its gradual disappearance.[40] The anthracological analysis of samples of charcoal has not only confirmed that deciduous trees existed in the area in the past, but it also helped with the identification of species (see Appendix). Data on paleosols at the Donets hillfort (within the modern city of Kharkiv) show that the stronghold was located on a promontory, with a forest immediately to the west, which survived well into the 13th century.[41] The Donets hillfort appears to have been located on the boundary between the steppe and the forest-steppe belts in the region of the Donets river.

The growth of the forest in of the forest-steppe belt during the late Holocene was the result of increased humidity during the late Bronze Age, and then by the relative stabilization of climatic conditions during Subatlantic phase of the Holocene, the beginning of which coincided with the early Iron Age. Between 2,800 and 2,500 years ago, therefore, the steppe gave way to the deciduous forest.[42]

The Subatlantic period of the Holocene had several climate phases. The available data show that the historical period considered in this book (mid-8th to mid-10th century) was a relatively arid episode, marked by the clarification of arboreal vegetation in the woodlands, forest glades, and the increased role of grasses in soil formation under forests. Degradation of forest vegetation and its replacement by steppe landscape most likely took place during the Medieval Warm Period in the forest-steppe zone, especially on forest edges (in areas of contacts between the forest-steppe and the steppe belts).[43] The productivity of the phytocoenosis in open meadow-steppe areas of the forest-steppe belt decreased considerably in the 8th to 10th centuries, thus making room for drought-resistant herbaceous plants. Moreover, studies conducted at the Koltunovka hillfort, on the river Tikhaia Sosna river (region of Belgorod, Russia) indicate that the share of cereals in palaeobotanical samples increased during the period of aridity coinciding with the Saltiv culture.[44]

Most scholars studying the history of the environment in the southern parts of Eastern Europe have noted the general expansion of forests and the displacement of the southern boundaries of natural zones to

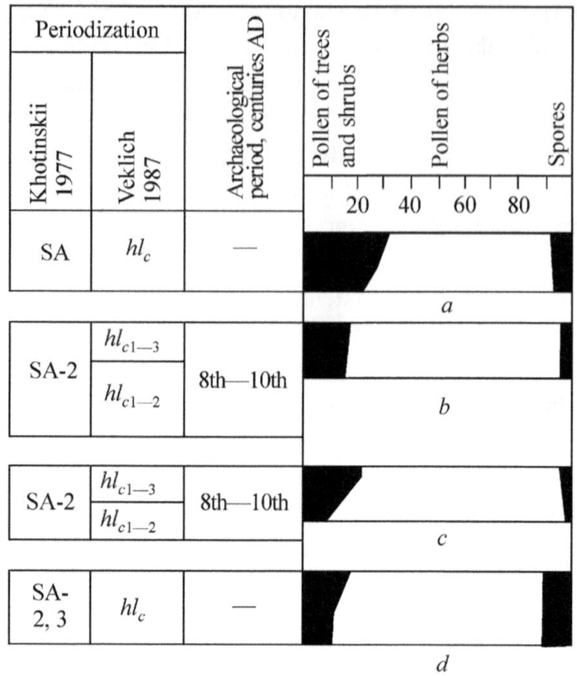

DIAGRAM 3.1
Spore-pollen curves of deposits from different archaeological sites: *a* — Glyboka 2; *b* — Komyshuvata XIV; *c* — Komyshuvata XIX; *d* — Amvrosiivka (after Matviishyna 2010, fig. 4.5, 4.6; made by S. A. Gorbanenko)

the south. According to palynological analyses made at Glyboka 2, Komyshuvata XIV and XIX, and Amvrosiivka (Diagram 3.1), the increasing role of arboreal vegetation due to the reduction of herbaceous groups falls within the period SA-2 ($hl_{c1-2,\,3}$). That period was characterized by the mesophytication of the steppe, the expansion of the forest but with the low participation of broadleaf species, and the dissemination of alder, birch and pine. Those traits continued into SA-2B.[45]

Additional data on the history of vegetation came from the pollen analysis of samples taken from the Lopans'ke marsh (region of Kharkiv, Ukraine) For SA-2, researchers have noted the presence and gradual increase of anthropochores (particularly pollen of cereals and weeds that thrive in cultivated fields). Judging by those samples, areas near marshes were used for grazing animals, but cereals were cultivated on the margins as well.[46] In our opinion, the only way that cultivation could have happen was by reducing groundwater and draining the marshes (see above). In turn, this was the result of the dramatic reduction of annual precipitation levels, which fed marshes — in short by increased aridity.

39 Goleusov et al. 2002.
40 Matviishyna, Karmazynenko, Kvitkovs'kyi, and Zadverniuk 2012, 250–52.
41 Koloda, Lisetskii, and Chendev 2002–2003, 166.
42 Aleksandrovskii 2002.
43 Aleksandrovskii, Chendev, and Trubitsyn 2011.
44 Chendev and Sarapulkin 2008; Chendev and Sarapulkin 2012, 99.
45 Matviishyna 2010, 164 figs. 4.5 and 4.6. The periodization used in the text is not that of the original study.
46 Bezusko, Bezusko, and Grechishkina 2008, 339–341.

In conclusion, against the background of an arid climate, forest clearing took place in the last quarter of the first millennium in the contact zone between the Slavs and the forest-steppe population of Khazaria. This created pastureland for livestock near settlements. On the other hand, increasing grass under forest cover and the appearance of forest glades with a microclimate that caused high herbage contributed to the expansion of forest lands suitable for grazing and hay. The reduction of water flow in rivers as a result of aridity led to a general lowering of groundwater, natural drainage, and fewer floods in floodplains. All of that influenced favorably the natural increase in land fertility, which made possible the use of those areas for agriculture, particularly soil cultivation and horticulture.

1.6 *Soils*

The general characteristics of relief in the region under study dictate the distribution of soils (Map 3.1). Zonal soils (appropriate to increased watershed positions) are gray forest soils formed under deciduous forest type of vegetation and forest-steppe subtypes of chernozem, many of which (except podzolized chernozems) were formed under herbaceous vegetation that grows in meadows between rivers.[47]

The zonal soils in the region formed no more than 4,000 years ago, primarily because of the steppe zonal soils signaling that what is now the forest-steppe belt in the center of the Eastern Europe used to be steppe land.[48]

Gray forest soils that were formed under watershed deciduous forests evolved from the steppe chernozems of the middle Holocene, as forests began expanding into the steppe lands during the late Holocene. The origin of gray forest soils in the Eastern European forest-steppe zone, as deriving from the chernozems of the Middle Holocene, has been discussed in several studies.[49] Among all gray forest soils in the territory under study, the dark gray forest soils are the most fertile.[50] Those soils, which appear in place where the forest has been cleared, are now widely used for plowing.

The second zonal component of the soils in the forest-steppe belt are the fertile chernozems, which are of three types (in the order of increasing fertility): podzolized, leached and typical.[51] Chernozems were formed mainly under herbaceous vegetation, except the podzolized variety, which represents the transition from meadow-steppe chernozems to the gray forest soils after the expansion of the forest cover. Their fertility is directly proportional to the number of plant residues, falling into soil, thereby increasing the capacity of useful chemical elements.[52]

Soils that were formed under forest vegetation are sporadically distributed, although they account for about a quarter of the territory under study. Chernozems appear all over that territory, due to its geographical location to the south of the forest zone. Azonal soils that were formed in locations with an abundance of water (near river beds, in the lower reaches of ravines and in marshes) cover a small area, and represent no more than 10 percent of the total area. At the same time, although they are sporadically distributed, when such soils appear near rivers (next to ancient settlements) it is likely that they were the target of intensive use in the past. There are in fact almost no soils unsuitable for agriculture in the region.

Soils changed very little, if at all, during the period under study.[53] There are in fact no examples of chernozem transitioning to soils of the forest type (or vice versa) for the period between the 8th and the 10th century. In fact, soils change much slower than any other natural components (definitely not as fast or as often as temperature, precipitation, and vegetation).

In short, the early medieval population in the area under study had optimal conditions for the development of agriculture.

2 Settlements

The archaeological complex at Verkhnyi Saltiv archaeological complex is situated on the right bank of the Donets, which is cut by ravines and hollows, some of them with streams. The bank is steep and tall, between 100 and 200 m above the sea level (the highest elevation in the surroundings of the complex is 201.9 m above sea level, see Map 3.2).

The hillfort occupies a promontory flanked by deep ravines, 35 m above the modern Pechenega Reservoir (134 m above the average sea level). The settlement occupying the natural terraces is located on at least two levels, one at 100–105, the other at 120–125 m above sea level. Some parts of the ancient settlement are located in the river valley (95 m above sea level), but are now under the water.

47 See Vernander and Tiutiunnik 1986.
48 Aleksandrovskii and Aleksandrovskaia 2005, 106; Chendev 2008, 122.
49 Vernander 1963, 164–182; Aleksandrovskii and Aleksandrovskaia 2005, 105–110.
50 Efimova 1977.
51 Krupskii and Polupan 1979, 39–118.
52 Tikhonenko 1995, 8.
53 Aleksandrovskii and Aleksandrovskaia 2005; Chendev 2008.

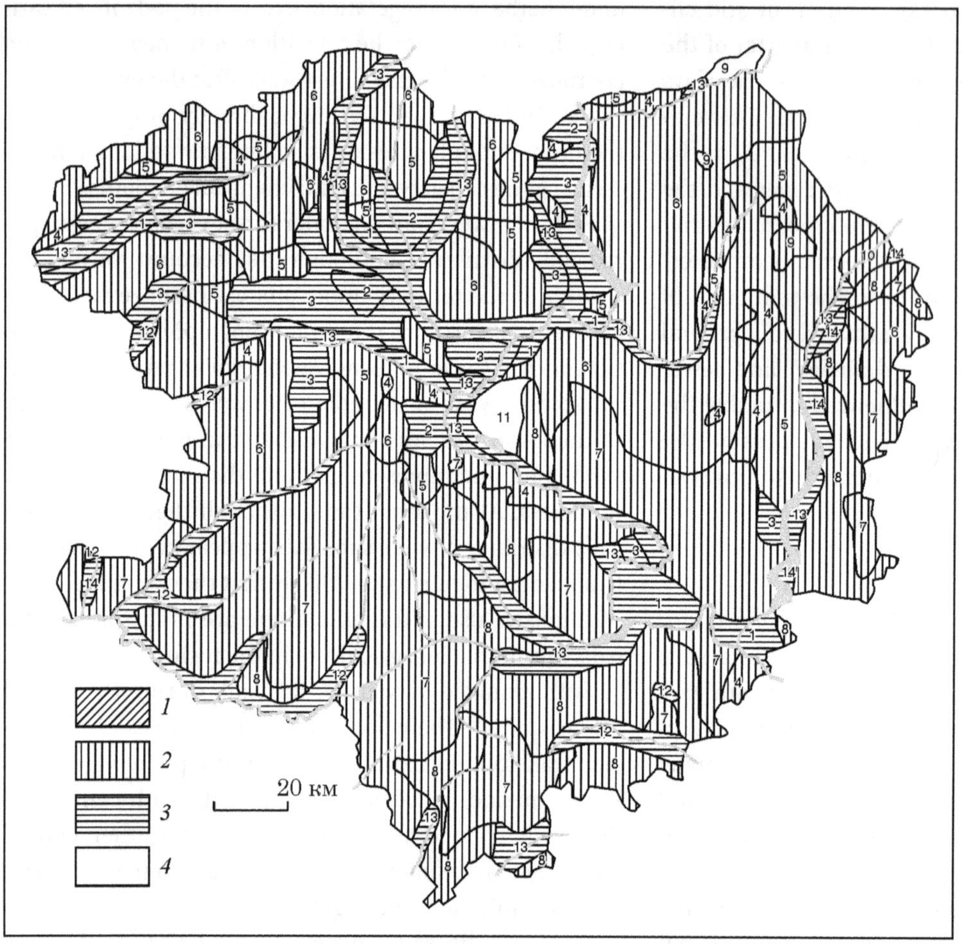

MAP 3.1 Soils of the Kharkiv region: *1* — soils formed under forest vegetation; *2* — chernozems; *3* — soils, the formation and fertility of which is affected by the humidity of the area; *4* — soils unsuitable for agriculture (based on Babychev and Marynych 1993, vol. 3, map "Kharkivska oblast. Map of soils"; made by S. A. Gorbanenko). According to the modern nomenclature, numbers correspond to the following types: *sod-podzolic soils on old alluvial and water-glacial sediments and on moraine*: 1 — sod-slight podzolic sandy, and clay and sandy soils combined mainly with their gleying types and peat-marsh soils; *gray forest soils and podzolized soils (soils non-washed off and washed off) mainly on loessial grounds*: 2 — light gray and gray forest soils, 3 — dark gray podzolized soils mainly on loessial grounds, 4 — podzolized chernozems and dark gray podzolized mainly on loessial grounds; *degraded soils (soils non-washed off and washed off) mainly on loessial grounds*: 5 — chernozems degraded mainly on loessial grounds; *typical chernozems (soils non-washed of and washed off) on loessial grounds*: 6 — mid-humused; *typical chernozems (soils non-washed off and washed off) on loessial grounds*: 7 — deep light- and mid-humused, 8 — mid-humused; *chernozems on heavy clays*: 9 — mainly solonetsous on heavy clays; *chernozems and sod gravelly soils on eluvium of dense ground*: 10 — chernozems and sod-carbonate soils on eluvium of carbonate rocks (marl, chalk, limestones); *solonetsous chernozems on loessial grounds*: 11 — solonetsous soils; *meadow chernozemic soils mainly on loessial grounds*: 12 — meadow chernozemic soils; *meadow soils on deluvial and alluvial sediments*: 13 — meadow soils; *sod mainly gleyed soils*: 35 — mainly gleyed sandy, clay and sand, and sandy loam soils together with slightly humused sands

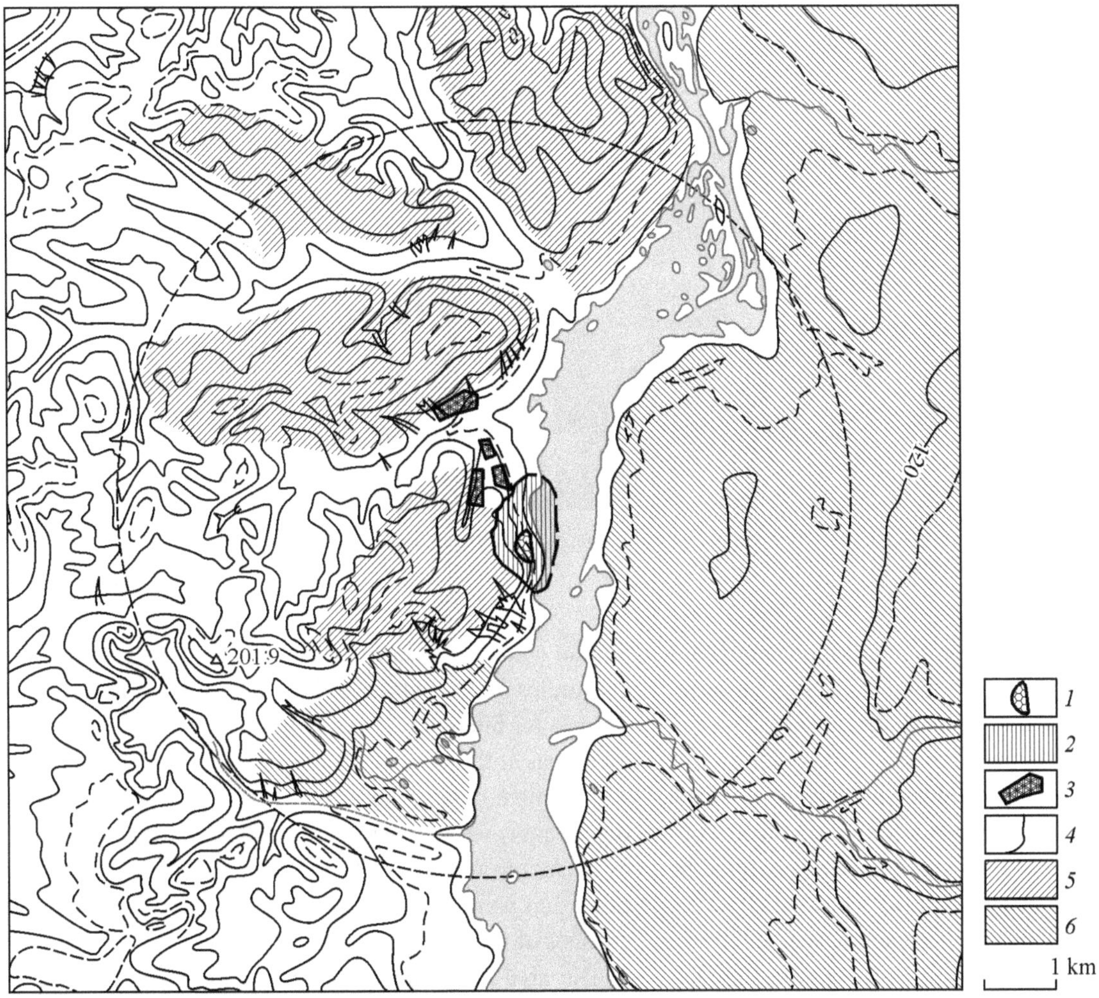

MAP 3.2 Verkhnyi Saltiv: *1* — hillfort; *2* — bailey (*posad*); *3* — cemeteries; *4* — elevation lines (every 20 m, after the Map of Ukraine, M-37–062); *5* — areas with relatively flat surface (suitable for soil cultivation); *6* — lands suitable for grazing livestock and haymaking (made by S. A. Gorbanenko)

A catchment area analysis (within a 5-kilometer radius around the settlement) shows that all available soils are good for agriculture. On the right bank, around the complex, one can find dark gray podzolized soils predominantly on loessial rocks. Those are soils that were formed under forest vegetation. Paleosols, buried under the medieval rampart, belong to podzolized chernozems formed under deciduous forest.[54]

There are two kinds of soils on the left bank: those whose fertility depends upon the proximity to water (the northern part), and chernozems (the southern part). The former are sod-weakly podzolic — sandy and clay-and-sandy soils in the complex with their predominantly gleying types, but peat-marsh soils on ancient alluvial and water-glacial sediments, and on the moraine. Chernozems are divided into podzolized and dark gray podzolized; they were formed predominantly on loess rocks.

The right-bank area was most likely reserved for fields. In fact, forests mark areas characterized by complex terrain with ravines and gullies also occurring sporadically within the potential resource zone, and behind it. They account for less than a half of the 5-kilometer catchment area (Map 3.2).

Because of strong anthropogenic, recent interference (the Pechenega Reservoir), it is rather difficult to single out floodplains in the river area. It is better to consider the riverbed of the Donets as a whole. The river has many oxbows, meanders and branches. The left bank is usually flat and low, and therefore periodically flooded (Photo 3.1). This is hardly observed now in the vicinity of the Pechenega Reservoir. Obviously, during the construction of the Pechenega Reservoir it was these areas that first fell into the flooded zone due to their low altitude. The lowland on the right bank has now completely disappeared, and cannot be observed of the left bank either. The recent afforestation targeted the first floodplain terrace (see Map 3.2).

54 Goleusov, Koloda, Lysets'kyi, and Chendev 2002.

PHOTO 3.1 Verkhnyi Saltiv — view from the hillfort to the left bank. Photo 1959–1961, expedition headed by D. T. Berezovets

The left bank of the 5-kilometer catchment area has a small flat section with favorable conditions both for farming and for animal husbandry (hayfields) (see Map 3.2).

On the basis of a careful examination of the landscape features and the photographs of the modern state of the surroundings, land plots may be identified that were conducive to agriculture. The only area partially separated by natural boundaries (riverbed, ravines) is located about 2.5 km to the south from the complex. Flat areas of small size appear near the settlement, to the north; about 1.5 km behind a distant burial ground, to the north-west; and about 3 km along the bank, up streams, to the north.

P'iatnyts'ke I is a settlement in the wide floodplain and on the sand dunes of left bank of the Velyka Babka river, a right-hand tributary of the Donets (map 3.3). The site extends along the river as a broad strip (about 1 km wide) for 2 km. The river is not difficult to cross, and territories on the opposite (right) bank mut have been easily accessible to locals. In the potential resource zone the altitude varies about 100 m (between 90.6 and 198.5 m above sea level). However, there are no sharp differences, and the adjacent territory of the left bank has a smooth slope towards the river (from east to west). Consequently, the adjacent areas are quite suitable for soil cultivation. In addition, the lands of the adjacent area were most likely flooded during the spring, which greatly contributed to the restoration of soil fertility (Map 3.3).

A part of the potential resource area is now covered by forest (Map 3.3, Photo 3.2). Paleopedological studies at Mokhnach, Verkhnyi Saltiv, and Donets'ke indicate an expansion of the forest during the 8th to 10th centuries.[55] Similar conclusions have been drawn from paleopedological studies in P'iatnyts'ke and in the vicinity. It is therefore not too far-fetched to assume that the area of the settlement was at the time at least as forested as it is now.

The meadow-chernozem soil that is predominant in the area formed in two stages of soil formation, separated by lenses of sandy loam. The old soil was formed in conditions that were more humid than today. It is clayey, and consists of meadow gley soil.

The area under study is a combination of landscape types that formed in the same climatic conditions: on the hills and on high slopes, deciduous forests with gray and dark gray forest soils on the hills and high slopes; on terraces above the floodplain, the forest-steppe (fragments of deciduous forests) with podzolized and leached chernozems; between rivers, the meadow-steppe with typical chernozems and meadow-chernozem soils.

In the river valleys, meadow and marsh landscapes are common; they occupy relatively small areas. Typical chernozems are found in floodplains, along with sod and sod-alluvial soils.

Gray forest soils were formed in places where there were and still are deciduous forests; and typical chernozems were formed in the areas, where meadow-steppe vegetation grew and still grows now. In the latter case, for a long time a meadow-steppe vegetation replaced a forest. Sod soils were also common.[56]

The Mokhnach hillfort is located on a promontory on the right bank of the Donets, next to a widely branched ravine, which only 15 years ago still had a stream. At the foot of the hillfort there are still many natural sources, and

55 Svistun, Chendev 2002–2003, p. 130; Goleusov at al. 2002; Koloda, Lisetskii, Chendev 2004, p. 166.

56 Matviishyna, Karmazynenko, Kvitkovs'kyi, and Zadverniuk 2012, 250–252.

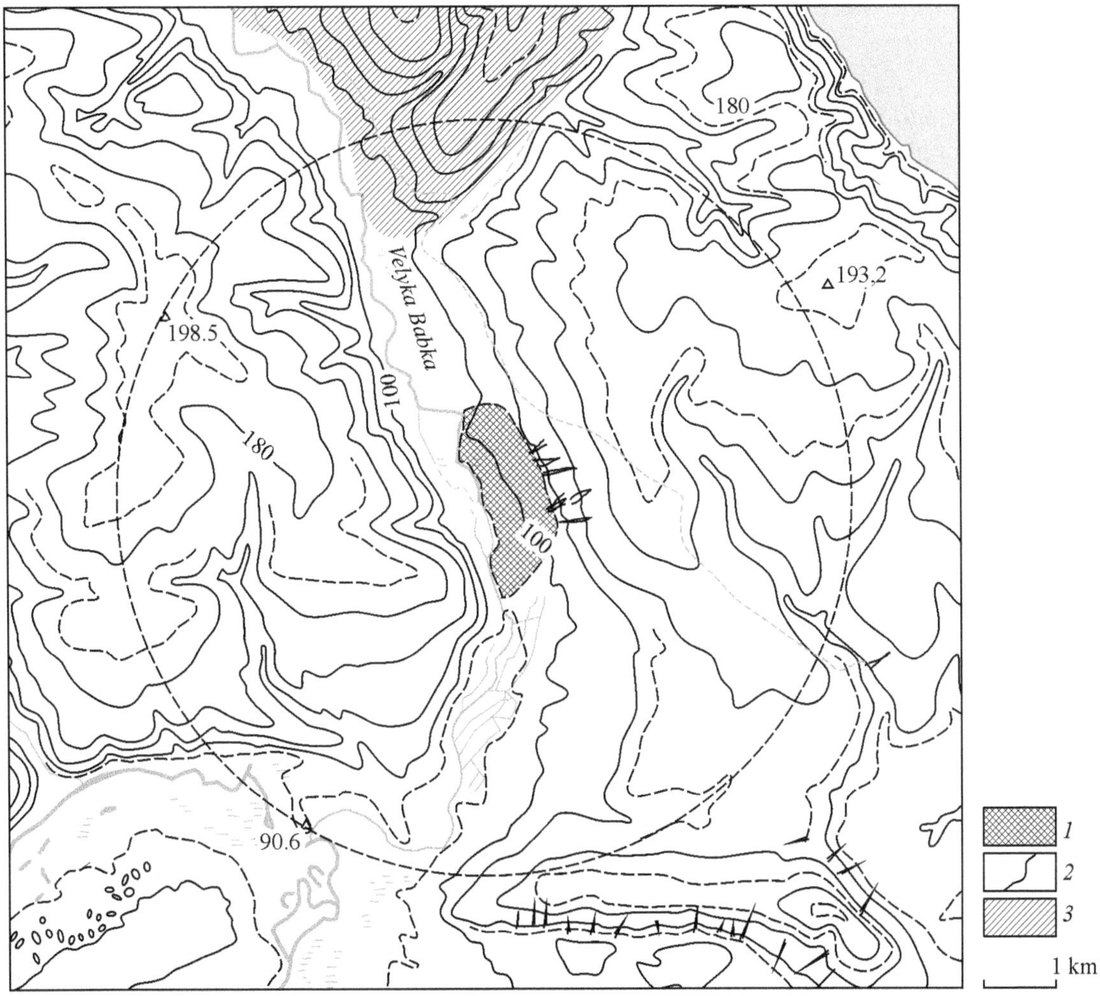

MAP 3.3 P'iatnyts'ke I: *1* — settlement (according to the data from the 2007–2008 excavations of V. I. Kvitkovs'kyi); *2* — elevation lines (in every 20 m, after the Map of Ukraine, M-37-074); *3* — lands suitable for grazing livestock and haymaking (made by S. A. Gorbanenko)

PHOTO 3.2
The valley of the Donets river near the P'iatnyts'ke I settlement
PHOTO BY V. V. KOLODA

according to the information of the old residents in the village, there were even more in the mid-20th century. The floodplain next to the hillfort has many oxbows, which has a great contribution to the generally moist terrain. Those are very favorable conditions for motley grass as the basis for hayfields, which is confirmed by the fact that the local population is engaged in mowing twice in the hot season. The relief around the site is irregular, dissected by ravines and depressions. Differences in elevations are no more than 80 m (between 100 and 180 m above sea level; with the highest elevation at 188.2 m; see Map 3.4).

An analysis of all the soils within a 5-kilometer radius around the site showed that the whole area around the hillfort could be used for agriculture. About 20 of the area in question has soils, the formation and fertility of which are constantly affected by proximity to water, while almost 80 percent consists of soils formed under forest vegetation.[57] Today almost all the entire territory is covered by forest. On the left bank, coniferous plantations prevail, with deciduous forest on the opposite bank (Photo 3.3).

Between 15 and 20 percent of the potential resource zone (with soils, the formation and fertility of which are constantly affected by proximity to water) is within the floodplain of the Donets (see Map 3.4). The boundary between the raised area and the floodplain, which must have been marked by different vegetation, is clearly visible on satellite pictures.

Near the hillfort, to the north, is a large area directly adjacent to the site. This was a quite suitable for farming. With a triangular shape, the areas situated on a raised plateau between depressions and ravines, directly behind the fortifications of the hillfort. Another similar area is located to the west from the hillfort, on a raised plateau between depressions and ravines. The distance from the hillfort to that location is about 500 m, as the crow flies. Cultivated fields now cover the southernmost part of the site (see Map 3.4).

About 800 m from the Donets is another large, flay area. That is where the Saltiv settlement is situated. No less than 17 Saltiv settlement sites have so far been identified around the hillfort in Mokhnach, along the Donets and its left-hand tributary, the Hnylytsa. It is therefore likely that the inhabitants of the Mokhnach hillfort or the residents of the nearest settlements of the Khazar age made extensive us of the floodplain.

One of the numerous satellite settlements near Mokhnach was Mokhnach P, which seems to have operated as the manufacturing center of the whole hinterland of Mokhnach microregion during the existence of the Khazar Khaganate.[58] It is located on the northern outskirts of the potential resource zone of the Mokhnach hillfort (Map 3.4; also see Map 1.6), and occupies three areas cut by ravines on the gently sloping edge of the plateau on the right bank of the Donets.[59] Despite a considerable development of ironworking, pottery production, and stone processing (rotary millstones making), the inhabitants of Mokhnach P were also engaged in agriculture for their own needs. An area suitable for agriculture is located directly to the west of the settlement, while the grazing fields nearest to the bend of the Donets towards the south-east.

Almost the entire potential resource zone at Koropovi Khutory has soils formed under forest vegetation.[60] The site is situated in the Hotvald (Zmiiv) forest, within the Homilsha National Park (Map 3.5). In addition to being covered by woods, the site is dissected by a right branch of the Donets, which has created gullies and ravines. Nevertheless, there are several areas with a sufficiently flat surface. One of them is to northwest from the settlement, most of it outside the modern forest. Another area is located to the south-west from the settlement, very close to the site.

The riverbed of the Donets and the surrounding area have a number of peculiar features (Photo 3.4). A satellite photo (see Map 3.5) shows that this is a band of about 1 to 1.5 km in width, with oxbows. It is impossible to establish the exact trajectory of the Donets during the existence of the Saltiv culture, but several sites located on its presumed banks can give a fair idea about where the riverbed was. There are several flat parts of the riverbank suited for agriculture. Such areas, enclosed by water from one side, and by terrain features from the other side, were good for pasturable or transhumant forms of grazing.[61] If the riverbed in ancient times had a different configuration, it is possible, that such or similar areas were present in the potential resource zone of sites in another localities.

During the excavations carried out in 2003–2004 by the expedition of "H. S. Skovoroda" KhNPU headed by Volodymyr Koloda, ancient roads and the place of a pier have been found.[62] The existence of the latter and the possibility of crossing to the left bank of the Donets suggest relatively easy access to the lowland areas in the floodplain. The local Saltiv community used those areas for animal husbandry, as well as soil cultivation.

57 Svistun and Chendev 2002–2003, 130.

58 Koloda and Koloda 2005–2009; Koloda 2010.

59 The current name of the site is "Midna," which means "copper."

60 Gorbanenko 2007, plate 4, 194.

61 Buniatian 1997.

62 Koloda 2008, 75.

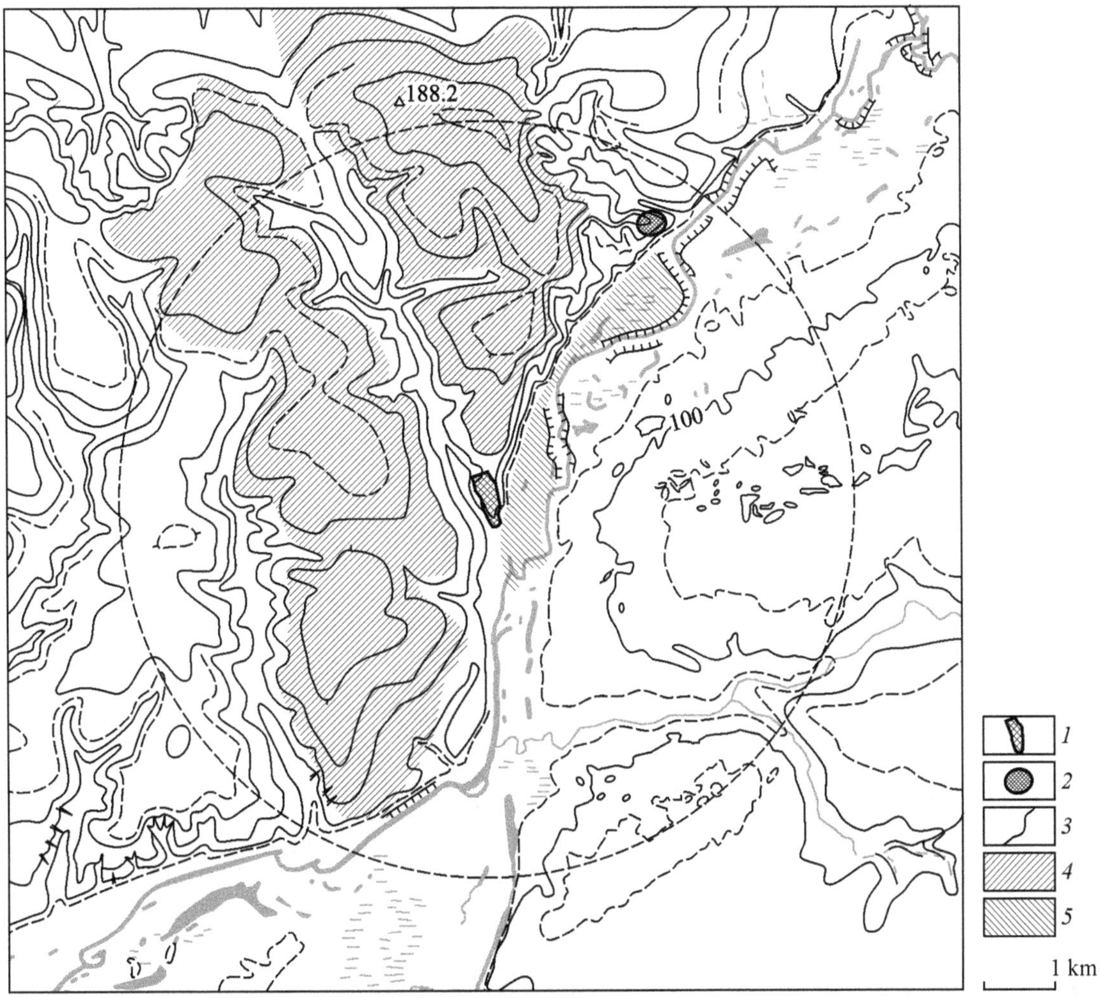

MAP 3.4 Mokhnach: *1* — hillfort; *2* — settlement P; *3* — elevation lines (in every 20 m, after the Map of Ukraine, M-37-073–M-37-074); *4* — areas with relatively flat surface (suitable for soil cultivation); *5* — lands suitable for grazing livestock and haymaking (made by S. A. Gorbanenko)

PHOTO 3.3 The valley of the Donets river near the Mokhnach hillfort
PHOTO BY D. IU. IUSHKOV, FROM PARAPLANE

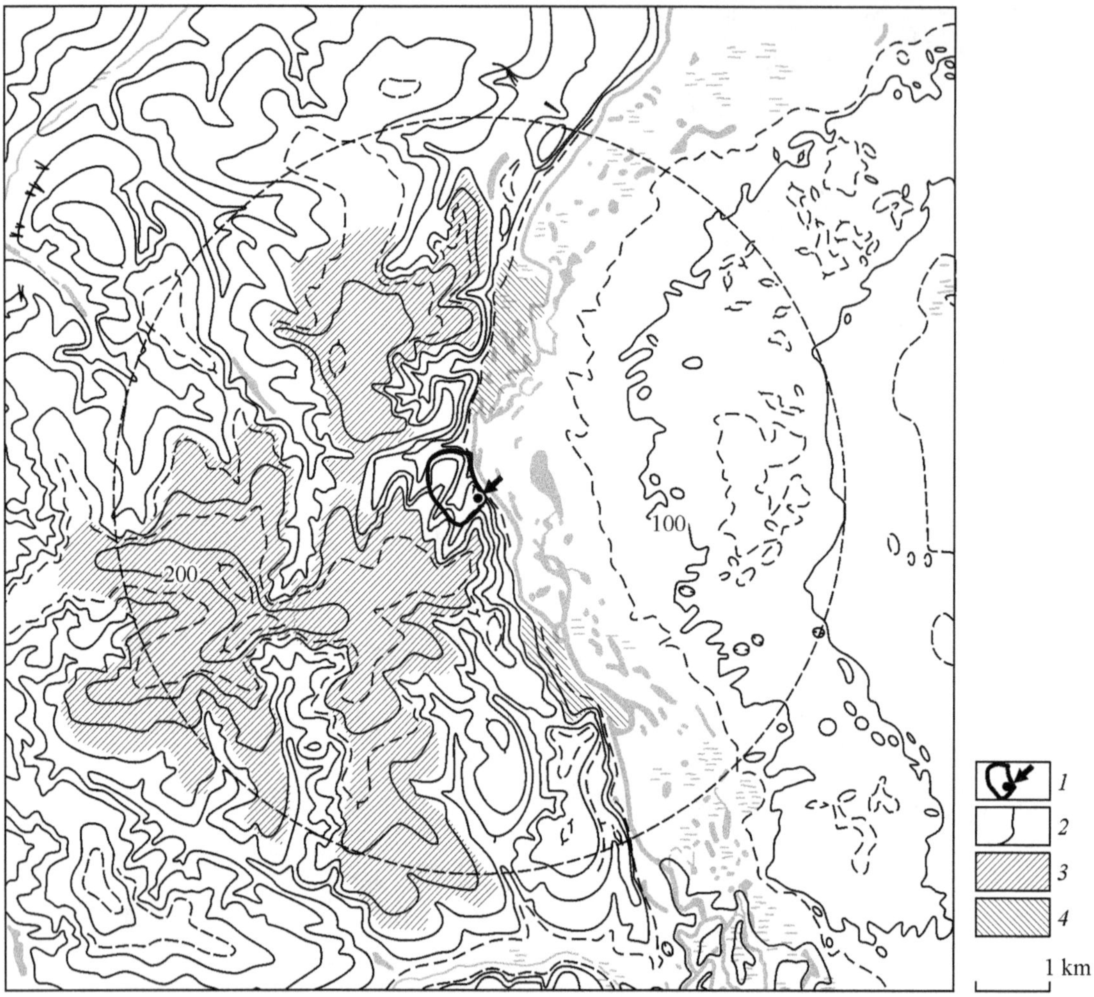

MAP 3.5 Koropovi Khutory: *1* — hillfort and settlement; *2* — elevation lines (in every 20 m, after the Map of Ukraine, M-37-085); *3* — areas with relatively flat surface (suitable for soil cultivation); *4* — lands suitable for grazing livestock and haymaking (made by S. A. Gorbanenko)

PHOTO 3.4
The valley of the Donets river near Koropovi Khutory
PHOTO BY V. V. KOLODA

The analysis of the zones around the sites under study in this book has showed the presence of areas suitable for soil cultivation. These areas have both common and different features. The undoubtedly most common feature is the location in the floodplain. The old pathways in the outskirts of the Koropovi Khutory hillfort, as well as the pier suggest easy access to the floodplain from the left bank of the river.[63] Such strategies are known from elsewhere in Eastern Europe.[64]

Lands in floodplains are most convenient for agriculture, as their cultivation required minimal effort. Some have pointed out that the floodplain land does not need to rest and regenerate, as those lands were periodically renewed by means of floods.[65] Therefore, there was no limit to the use of floodplains and terraces above them. However, with the development of farming, which opened the possibility of cultivating other areas, as well as the introduction of advanced techniques of animal breeding, the floodplain ceased to be used for crop cultivation and was turned instead to cattle breeding. Several rural areas maintain that specific use of floodplains.

The potential zones around three settlements — Mokhnach, Koropovi Khutory and P'iatnyts'ke I — included forested areas.[66] The beginning of agricultural activity must have post-dated ample clearing works. It is quite possible that initially swidden cultivation was practiced, with clearings ready for the plow.

Be that as it may, the inhabitants of Verkhnyi Saltiv practice arable farming from the beginning of their settlement.

Two complexes (Verkhnyi Saltiv and Koropovi Khutory) are quite large, and must have been well populated.[67] Such sites as Chuhuiv and Mokhnach, both hillforts that operated as local centers, were surrounded by smaller settlements. Some of them, such as Mokhnach P, were service settlements specializing in craft activity. In other words, the population of those service settlements needed to be fed from the outside.

The study of the environment of archaeological sites reveals the presence of excellent natural conditions for agriculture. In all known cases, soils suitable for farming were distributed on fairly flat areas. The presence of floodplain areas mainly used for grazing supplements the picture.

63 Koloda 2008, 75.

64 Voevodskii 1949, 74; Karavaiko and Gorbanenko 2012, 25.

65 Krasnov 1967, 20.

66 Svistun and Chendev 2002–2003, 130; Chendev and Koloda 2012; Matviishyna, Karmazynenko, Kvitkovs'kyi, and Zadverniuk 2012.

67 This assumption can be verified only through future large-scale excavations.

CHAPTER 4

Arable Farming

To analyze agriculture in detail, we have divided the materials into three groups on the basis of production processes agricultural tools, palaeobotanical data, as well as the collection, processing and storing of crops. The general characteristics of agriculture in the forest-steppe region at the time of the Saltiv culture were discussed by Volodymyr Mikheev.[1] Our goal in this chapter is to offer a complex analysis of arable farming in the region, which deals not only with tools, but with natural conditions and palaeobotanical data as well.

The sites taken into consideration for that analysis are Verkhnyi Saltiv, Mokhnach (both stronghold and the settlement called Mokhnach P), Koropovi Khutory, and P'iatnyts'ke I settlement. In addition, we referred to other sites in the forest-steppe zone of the Donets basin, particularly for finds of agricultural implements and for palaeobotanical data. All sites produced evidence of agricultural, such as plowshares and coulters (for soil tillage using draft animals), hoe-heads and spade frames (for soil tillage using human power, or for secondary soil tillage and digging for non-agricultural purposes); sickles (for harvesting) and scythes (for producing fodder for livestock, possibly for harvesting as well); millstones and graters. Palaeobotanical data has been collected from Verkhnyi Saltiv, Mokhnach (stronghold), Koropovi Khutory, P'iatnyts'ke I, Chuhuiv, and Mokhnach P.[2]

1 Tools for Primary Soil Tillage with the Use of Draft Animals

The tools for primary soil tillage with the use of draft animals are plowshares and coulters. The former (Plate 4.1) are of type I B 2, with a blade wider than the socket (Plate 4.1, *1–8, 12*). Verkhnyi Saltiv, the stronghold in Mokhnach and the satellite settlement at Mokhnach P have produced two specimens each, but a pair of plowshares is also known from the Koropovi Khutory hoard.[3] Another plowshare of the same type was found in P'iatnyts'ke I. Narrowly bladed plowshares of a different type (I B 1) are known from Petrivs'ke.[4] A fragment of a plowshare, possibly narrow-bladed as well, was found in Mokhnach P. The same site produced another fragment, a socket of a plowshare of unknown type. Some plowshares were reinforced with welded metal bands (Plate 4.1, *6–8*), which may indicate repair of such important arable tools, or an attempt to prevent the premature wear of the tool.

Plowshares come in different sizes. The largest is one of the two specimens from Verkhnyi Saltiv — 30 cm long (the length of the blade is 21.6 cm), and 16.8 cm wide. The smallest is the plowshare from P'iatnyts'ke I — 14 cm long (the length of the blade is 9 cm) and 9 cm wide. The length of the socket correlates with the general size of the tool, but its width depends upon the size of the wooden base to which the plowshare was attached. That plowshares varied so widely in size is most likely the result of their use for variable soil quality, crop selection, and draft power. Small plowshare were suitable for virgin soil — the first plowing after land clearing, in floodplains, and especially for planting millet. By contrast, massive plowshares worked very well in old fields, as deep plowing was the best condition for higher yields for the dominant crops (but not for proso millet).

Plowshares from sites in the forest-steppe zone of Khazaria have good analogies elsewhere in the Khaganate — at Tsimliansk (a hillfort in Right-Bank Ukraine), at Mayaki, Sydorove,[5] and Gosudarev Iar.[6] Most belong to type I B 2, which may be interpreted as evidence that they were used for plowing land recently cleared from shrubs or trees.

Coulters have been found both whole and in fragments (Plate 4.2). Two specimens are known from Mokhnach P, two more from Mokhnach Zh[7] and P'iatnyts'ke I, respectively. Three fragments have been found inside the

1 Mikheev 1985, 32–52.

2 The data acquisition from all those settlements was conducted by Serhiy Gorbanenko. The analysis of imprints on clay products from the first four sites was done by Galina Pashkevich, and by Serhiy Gorbanenko for the other two sites. The materials from flotation in Mokhnach (2015) and P'iatnyts'ke I (2016) were identified by Serhiy Gorbanenko.

3 Besides the two plowshare, the hoard includes 13 stemmed sickles, all of the same type, discussed below (Koloda 2014b).

4 Mikheev 1985, fig. 22, *1*.

5 Pletneva 1994, fig. 46, *1, 2*; Mikheev 1985, fig. 22, *2–11*.

6 Davydenko and Grib 2011, fig. 2, *4*; 4; 5, *4*; 6, *1, 2*; 8, *5*; Koloda 2013, fig. 3; 5, *2*; 8; 10, *2*.

7 The index Zh of this settlement corresponds to Cyrillic letter Ж, which was used for designating another satellite settlement in the hinterland of the Mokhnach hillfort.

 DOI:10.1163/9789004429574_006

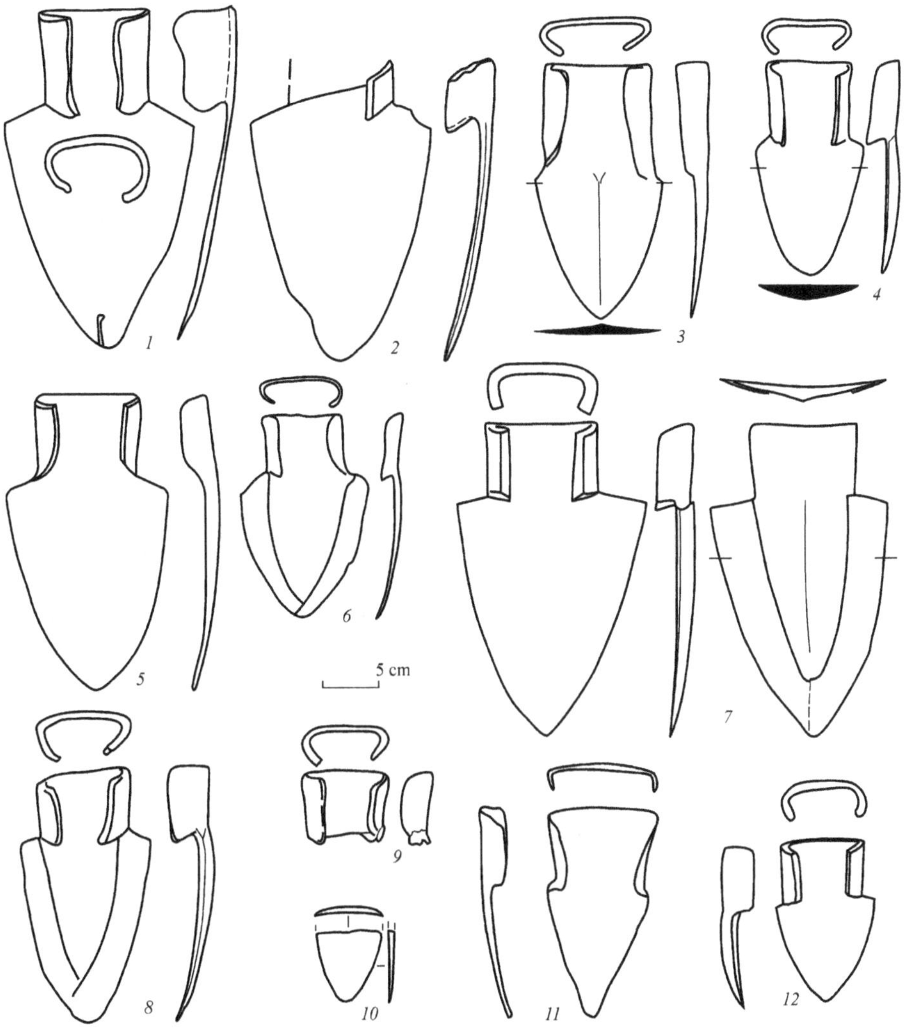

PLATE 4.1
Plowshares: *1, 2* — Verkhnyi Saltiv; *3, 4* — Koropovi Khutory; *5, 6* — Mokhnach; *7–10* — Mokhnach P; *11* — Petrivs'ke; *12* — P'iatnyts'ke I

stronghold in Mokhnach hillfort, another in Koropovi Khutory. According to V. G. Borodulin, the former director of Verkhnyi Saltiv museum, two coulters have been found on that site. Another coulter was found in 2005 in Sukha Gomil'sha (Plate 4.2, *10*).

Judging by the whole specimens, those were coulters are of stemmed type. The description and analogies indicated for the Verkhnyi Saltiv tools,[8] as well as for the fragments from Koropovi Khutory and the Mokhnach hillfort, those were also coulters of the stemmed type. On the basis of the current state of research on early medieval agricultural tools, especially those of the Saltiv culture, one can distinguish two types of coulters with different forms of the working part.[9] One of them has a feather-shaped working part. This type has a long rod and a relatively short, symmetrical, and double-edge blade. The second type includes coulters with a knife-shaped working part with only one edge. Bent towards the blade, the working part of the second type represents between a half and a third of the total length of the tool. The coulters from Sukha Gomil'sha and a stray find from Kochetok (near Kharkiv) belong to the type I (Plate 4.2, *2, 10*). By contrast, the specimens from Koropovi Khutory, the Mokhnach hillfort, Mokhnach P, Mokhnach Zh, and P'iatnyts'ke I are all of the second type (Plate 4.2, *1, 3–9*). It is important to note that so far no socketed coulter has been found on any Saltiv site, although such coulters were common on contemporaneous sites of the Eastern Slavs.[10]

Coulters found on Saltiv sites are between 39 and 45 cm long (with a blade varying between 13 and 15.5 cm), with stems between 3 and 3.3 cm thick.[11] There are, however, several differences between specimens. The tool from Sukha Gomil'sha, for example, has a rod on the side opposite to the blade and is bent at a right angle to the working part, forming a slightly narrowed, 5-cm long "spike," with

8 Mikheev 1985, 36–37, fig. 23, *3–7*.

9 Koloda 2013, 74 with n. 1.

10 Baran 2004, fig. 37, *6*; Gorbanenko 2007, pl. 6; fig. 9;10; Kovalevskii and Gorbanenko 2014.

11 Mikheev 1985, fig. 23, *4–7*.

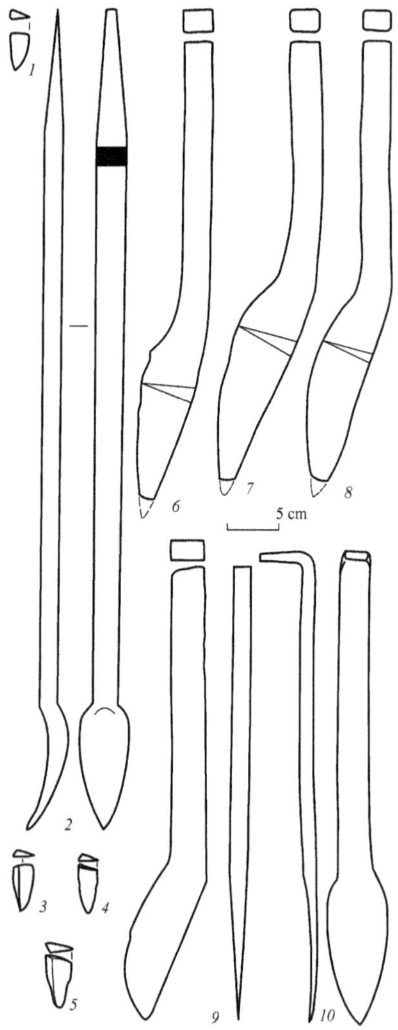

PLATE 4.2
Coulters: *1* — Koropovi Khutory;
2 — Kochetok (accidental find);
3–5 — Mokhnach; *6* — Mokhnach Zh;
7, 8 — Mokhnach P; *9* — P'iatnyts'ke I;
10 — Sukha Gomil'sha (accidental find)

which the coulter was attached to the wooden base of the tool. The blade itself has a spear-shaped (or leaf-shaped) form and is 13 by 6 cm (Plate 4.2, *10*). A similar tool was found in the Mayaki hillfort, but was wrongly interpreted as plowshare.[12]

The largest coulter was accidentally found near Kochetok (Plate 4.2, *2*). The stem of the tool, created from a rod with rectangular cross-section (2.7 by 1,6 cm) is 66.5 cm long. Its outer (upper) end, 12 cm long, is narrowed. The length of the feather-like working part is 10 cm, and its width is 5 cm. In all respects, this tool is almost identical to that found in Mayaki, while the size of its working part is close to that of the coulter from Sukha Gomil'sha.

Plowshares have commonly been found together with coulters, as in Tsimliansk, Mayaki, Tepsen', Gosudarev Iar, and Mokhnach P.[13] This confirms that the two tools were used in combination. This raises the question of why were iron details added to an ard made of wood.[14] One cannot simply exclude the possibility that wooden ards were used. Such ards had heads placed at an angle of about 45° (type *1*). Wooden body ards had heads placed horizontally, parallel to the ground (type *2*). At any rate, finds of iron parts suggest that diversity of forms. If the ard-head was used alone, without coulter, it was typically a body-ard with a plowshare placed horizontally, parallel to the ground (type *4*). If the ard-head was used together with a coulter, then that was a plow with a plowshare placed horizontally, parallel to the ground, behind the coulter. That plows most likely had a mouldboard as well (type *5*; see fig. 2.4).

The variation described above was most likely a response to land plots with different soils, including chernozems and areas cleared from forest vegetation. In addition, if one assumes that floodplains were used for soil cultivation, not cattle breeding, then a simple ard made completely of wood would have been sufficient — an ard with the ard-head placed at an angle of about 45° or an ard with a runner not fastened with iron details.

2 Auxiliary Tools for Soil Tillage

Two types of tools may be considered auxiliary in relation to soil tillage — hoes and spades. Spades come with either semicircular or rectangular blades. The former were often reinforced with an iron frame, such as found in Verkhnyi Saltiv, Vovchansk,[15] and the Mokhnach hillfort. Frames for rectangular blades have been found in the Mokhnach P settlement. Semicircular blades are much larger, between 18 and 23 cm in length, 16 to 20.5 cm wide. The metal reinforcement is a strip of 3 to 8 cm (Plate 4.3).

However, as A. A. Laptev has noted, all known semicircular spade blades come from sites that, in addition to an early medieval occupation phase, have been occupied

12 Mikheev 1985, p. 32, fig. 23, *3*.

13 Liapushkin 1958a, p. 117, fig. 10; Mikheev 1985, 131 and 132; Frondzhulo 1968, 144, fig. 10, *1*, *2*; Davydenko and Grib 2011, 251–253; Koloda 2013; Koloda 2015, 113.

14 It is important to note that, with the notable exception of peat bogs, the natural conditions in modern Ukraine are not conducive to the preservation of archaeological timber.

15 Mikheev 1985, fig. 24, *21*, *22*.

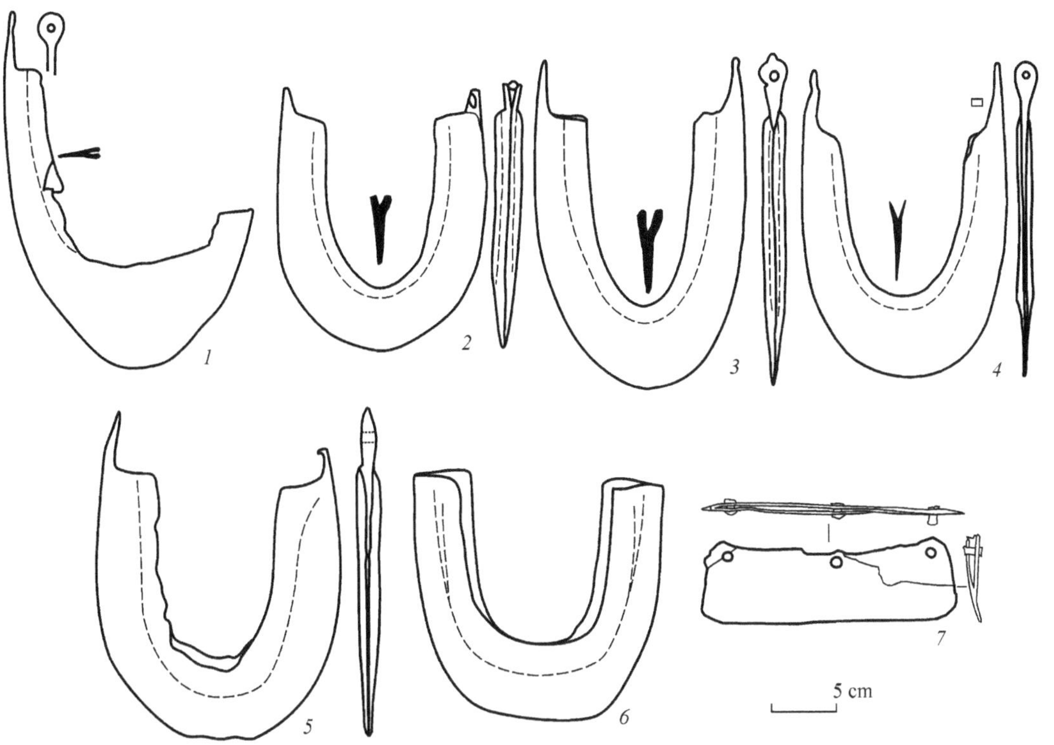

PLATE 4.3 Spade frames: *1–6* — rounded; *7* — rectangular (*1–3* — Verkhnyi Saltiv; *4, 5* — Vovchansk; *6* — Mokhnach S; *7* — Mokhnach P)

during the 17th and 18th centuries. None of them was found on a Saltiv site with only one phase of occupation (dated to the early Middle Ages).[16] So, only the rectangular blades may be considered as certainly typical for the Saltiv culture. This is further confirmed by two similar blades found during excavations in the Mayaki hillfort.[17] Each of them was about 10 cm, while the specimen found in Mokhnach P was twice as long. All the three rectangular spade blade had reinforcements made of two strips of iron (5 to 7 cm wide and 0.2 to 0.3 cm thick). The strips were welded on one side (working blade), and on the other side, where a wooden base of a spade was inserted, they were connected by two or three through rods (plate 4.3, *7*).

Numerous hoe-heads with a vertical socket are known from sites of the Saltiv culture (plate 4.4, *1–14*). They were also deposited in graves. They come from funerary sites.[18]

The number of such hoe-heads was so great in the catacombs of the Dmitrievka cemetery that Svetlana A. Pletneva could advance a classification with no less than 9 types. She interpreted that remarkable variation as an indication that hoes were made by different artisans, and not that the tool "evolved" from simple to complex.[19] Whatever the meaning of the hoes deposited in graves, they certainly suggest that they were primarily used to dig (grave) pits, not for farming.

Hoe-heads with vertical sockets seem to have had multiple functions. interesting in that in Saltiv cultural environment they were multifunctional. Small specimens with narrow blades were woodworking tools, while the larger specimens could very well have been used for farming purposes as well. In fact, the hoes found on settlement sites are larger than those found in graves, which suggests that they were used in farming.[20] Such are the specimens found on the Mokhnach P and Mokhnach U[21] settlement sites. The large hoe-heads with vertical sockets (which were presumably used in farming) are between 8 and 10.5 cm long,[22] with blades between 4.5 and 8.5 cm of width. Such hoes are well known from other Saltiv sites

16 Laptev 2007.
17 Mikheev 1985, fig. 19, *2, 3*.
18 Kukharenko 1952, 41; Mikheev 1985, 117–119; Pletneva 1989, fig. 46; Pletneva 1999, fig. 12; Aksenov 2012, fig. 4, *3*.

19 Pletneva 1989, 91–93.
20 A hoe found in a Saltiv catacomb is between 7 and 8 cm long, and 5 cm wide.
21 The index U of this settlement corresponds to Cyrillic index У, accepted in Ukrainian and Russian literature.
22 The exception is a tool from the Mokhnach U settlement, with a total length of more than 20 cm (this brings it closer to another group of the ard-heads with a horizontal socket).

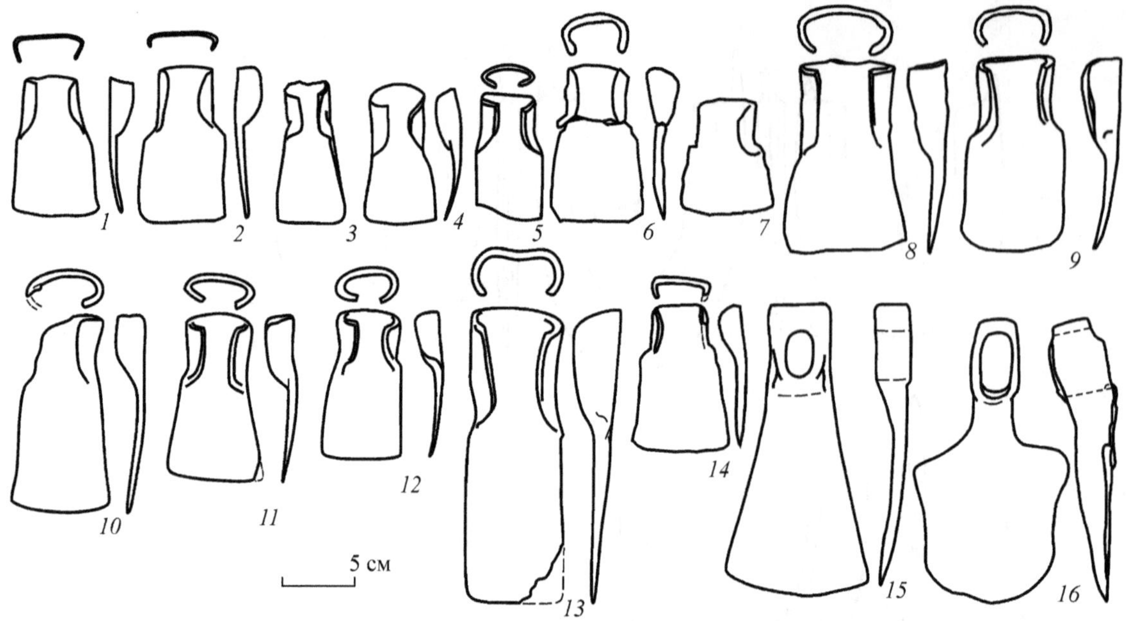

PLATE 4.4 Hoe-heads: *1–14* — with vertical socket; *15*, *16* — with horizontal socket (*1–4*, *15* — Verkhnyi Saltiv; *5*, *16* — Mokhnach; *6–11* — Mokhnach P; *12* — Mokhnach T; *13* — Mokhnach U; *14* — P'iatnyts'ke I)

the Sydorove hillfort, Sarkel, Tsimliansk, as well as the sites of Taurica and Heroivka 3 (Kerch Peninsula).[23]

Hoe-heads with horizontal, solid sockets (plate 4.4, *15*, *16*), although with similar blade parameters, have quite different silhouettes. Most have almost no neck separating the blade from the socket (see Plate 4.4, *15*). However, there are also specimens with clearly marked necks.[24] Such a hoe-head was found in the Mokhnach hillfort (plate 4.4, *16*). The hoe-heads with expressed neck and semicircular blade resemble Central Asian *ketmens* in shape.[25]

Plate 2.7 shows a general view (graphic reconstruction) of both varieties of hoes is represented.

Most important for our analysis are assemblages in which arable tools (plowshares and coulter) appear together with hoe-heads.[26] Sometimes both types of hoe-heads appear in hoards, which only confirms their joint use by the farmer.[27]

23 Kravchenko and Davydenko 2001, p. 249, fig. 43, *1–3*; Artamonov 1958, fig. 29, *1*; Sorokin 1959, p. 145, fig. 5, *1–8*; Liapushkin 1958a, fig. 10; Mikheev 1985, fig. 24, *1–8*; Ponomarev 2012, 67, fig. 2, *1*.

24 Mikheev 1985b, fig. 24, *8*.

25 *Ketmen* is a typically Central Asian hoe.

26 Koloda 2013, fig. 3; 5.

27 Davydenko and Grib 2011, 25.

3 Palaeobotanical Data

Among all imprints of grains on clay artifacts that have been identified, more than half are of proso millet (Diagram 4.1; Table 4.1). The rest are barley (with its bottle-shaped variety), emmer wheat, naked, common wheat, rye, and oats. Both peas and weeds are rare (Plate 4.5).[28]

In addition, palaeobotanical macro-residues have been recently obtained by flotation from two sites. On the hillfort site at Mokhnach, 10 to 20 cubic decimeter-large samples were obtained in 2015 from four features, as follows: fragments of hazelnut shell (*Corylus avellana*) from feature 79; 1 grain of proso millet, 1 of barley, 1 of rye, and 3 fragments of grains of unidentified cereals from feature 85; 1 grain of proso millet, 2 of emmer wheat, and 1 of rye from feature 88; and 2 grains of proso millet, 1 of rye, and 1 fragment of an unidentified species from feature 90.[29] A 50 cubic decimeter-large sample obtained in 2016 from pit 16 in P'iatnyts'ke I produced 3 grains of proso millet, 1 grain of rye, 3 fragments of grains of unidentified species, and 1 half of a pea. The results are unimpressive, and cannot be used for statistics. However, this is to be treated as a reliability test for flotation on sites of the Saltiv culture in the forest-steppe zone along the river Donets.

28 Gorbanenko 2013.

29 Gorbanenko 2016.

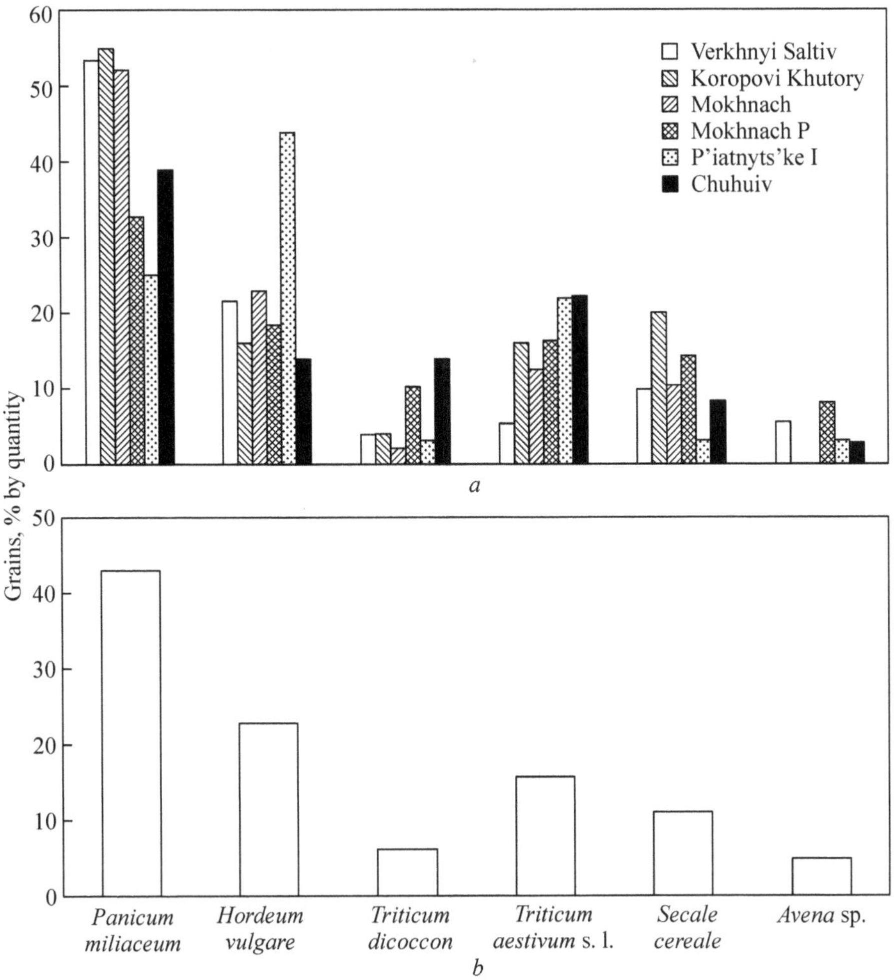

DIAGRAM 4.1 Palaeobotanical spectra of cereals from Saltiv sites in the region under study (*a*), and general palaeobotanical complex (*b*), by quantity

Among the imprints of cereal grains on clay artifacts, the most numerous are those of proso millet (*Panicum miliaceum*), its grains without glume, and its chaff. Proso millet grains are 1.4 to 2.2 by 2 to 2.9 mm. The imprints of grains without chaff are between 1.4 and 1.8 mm in diameter. Imprints of grains of barley (*Hordeum vulgare*) are between 3 and 4.5 mm wide, and between 7.5 and 9 mm long. The ratio between length and width is 2.3 to 2.5. A variety of barley known as *Hordeum vulgare* var. *lagunculiforme* has been documented in Verkhnyi Saltiv. Those grains are between 3 and 3.7 mm wide, and between 8.2 and 10 mm long. A characteristic feature of this kind of barley is a potential six-row, as the lateral spikelets in every triplet are not sessile, but they have stalks from 1 to 3 mm long. All spikelets in triplets are fertile around the ear; sometimes lateral spikes in the ear are acarpous.[30] The stalk length on the grains from Verkhnyi Saltiv is about 2.5 mm. We have also identified an imprint of a naked variety of barley (*Hordeum vulgare* var. *coeleste*).

Wheat imprints are also known from the Saltiv. They belong to three different species: hulled forms, such as emmer (*Triticum dicoccon*) and einkorn (*Triticum monococcum*), as well as naked, common wheat (*Triticum aestivum* s. l.). The grains of the hulled varieties are between 2.5 and 3.6 mm wide, and between 5.8 and 7.8 mm long for emmer; 3.2 by 5.8 mm for einkorn (only one imprint from Mokhnach P). Naked varieties are typically smaller: between 1.6 and 4 mm wide, and between 4.5 and 6.4 mm long (with a length/width ratio of between 1.5 and 2.4).

Rye (*Secale cereale*) has been identified on the basis of imprints of grains that are between 2 and 2.9 mm wide, and between 6 and 9 mm long (with a length/width ratio of between 2.6 and 3). Imprints of oats (*Avena* sp.) also were identified; they are between 2.2 and 3.2 mm wide, and between 8.2 and 9.2 mm long (with a length/width ratio of between 2.8 and 3.2). One a small number of

30 Bakhteeev 1956.

TABLE 4.1 Palaeobotanical materials from Saltiv sites of the region under study

Plant	Units of measurement	Sites					
		Verkhnyi Saltiv	Koropovi Khutory	Mokhnach	Mokhnach P	P'iatnyts'ke I	Chuhuiv
Panicum miliaceum L.	Quantity	60	11	25	16	8	14
	%	54.0	55.0	52.1	32.7	25.0	38.9
Hordeum vulgare var. *lagunculiforme*	Quantity	4	—	—	—	—	—
Hordeum vulgare L.	Quantity	20	4	11	9	14	5
	%	21.6	16.0	22.9	18.4	43.8	13.9
Hordeum vulgare var. *coeleste*	Quantity	1	—	—	—	—	—
Triticum dicoccon Shrank.	Quantity	4	1	1	5	1	5
	%	3.6	4.0	2.1	10.2	3.1	13.9
Triticum aestivum L.	Quantity	6	4	6	8	7	8
	%	5.4	16.0	12.5	16.3	21.9	22.2
Triticum monococcum L.	Quantity	—	—	—	1	—	—
Secale cereale L.	Quantity	11	5	5	7	1	3
	%	9.9	20.0	10.4	14.3	3.1	8.3
Avena sp.	Quantity	6	—	—	4	1	1
	%	5.4	—	—	8.1	3.1	2.8
Pisum sp.	Quantity	1	—	—	1	1	1
Bromus sp.	Quantity	1	3	4	3	—	3
Weed, not identified	Quantity	—	—	—	2	—	—
Ear, not identified	Quantity	—	—	—	—	—	1
Total	Quantity	114	28	52	56	33	41

Footnotes. In the "Units of measurement" column percentage is indicated only for major cereals; for all others percentage has not been counted (for Verkhnyi Saltiv, the percentage of hulled barley and "bottle-shaped" hulled barley are given together in the row for "*Hordeum vulgare* L.").

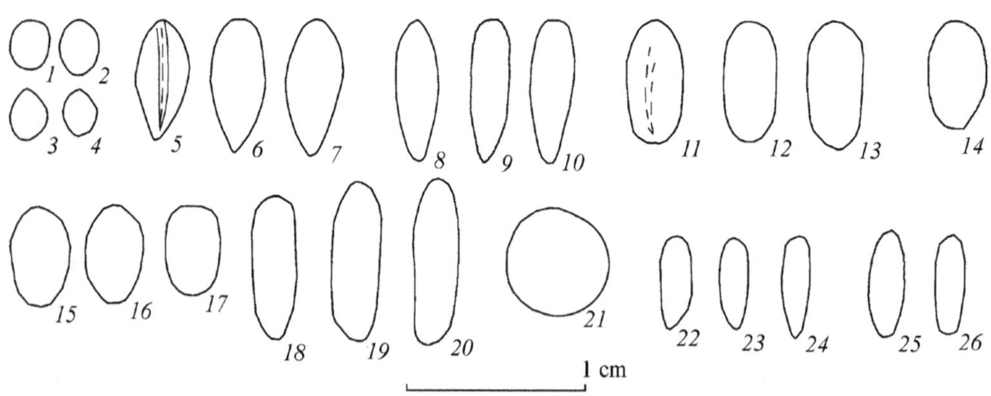

PLATE 4.5 Mokhnach P, imprints of grains on pottery: *1–4* — proso millet; *5–7* — hulled barley; *8–10* — rye; *11–13* — emmer wheat; *14* — einkorn wheat; *15–17* — common wheat naked; *18–20* — oats; *21* — pea; *22–24* — brome; *25, 26* — weed unidentified

imprints of peas (*Pisum sativum*) have been noted, with a diameter between 4 and 5.8 mm. The imprints of seeds of brome (*Bromus* sp.: *B. secalinus, B. arvensis*) are between 1.7 and 2.2 mm wide, and between 5.6 and 7.3 mm long (with a length/width ratio of between 2.5 and 3.8). For all species mentioned above, the size of actual grains from the sites in the region under study are about the same.

When compiling palaeobotanical spectra of the main cereals on the basis of their weight (see Chapter 2), some millet seed imprints were excluded (32 from Verkhnyi Saltiv, 13 from Mokhnach, all of those from Mokhnach P and P'iatnyts'ke I, 5 from Koropovi Khutory, and 3 from Chuhuiv (Diagram 4.2). A cluster analysis of the data showed the similarity of the results at 80 percent. The greatest similarity (87 percent) is between Mokhnach and Verkhnyi Saltiv, and the lowest (80.6 percent) is between Chuhuiv, Mokhnach P, Koropovi Khutory, P'iatnyts'ke I, Mokhnach (hillfort), and Verkhnyi Saltiv. The cluster analysis thus shows a sustained and elevated level of development of agriculture across generations, as well as the same adaptation to natural and climatic conditions.

When comparing cereal crops, similar and distinctive features are clearly visible. First, it is worth noting the small amount of proso millet in all PBS, between 5.7 and 9.6 percent. Oats are also represented by small shares, from 0 (in two cases) to no more than 7.2 percent. Between the hulled and the naked species of wheat, the latter prevail by at least 3 percent, and at most 24 percent. Emmer (hulled wheat) is represented by a minimum of 4.5 percent (in two cases), and a maximum of almost 25 percent (in one case), but in four other cases its presence is insignificant. The minimum percentage for the naked varieties of wheat is 11.8; another five indicators vary between a quarter and a third of the total mass of grains.

The indicators of rye are also unstable: the minimum percentage is 3.4, while the maximum is 30 (see Table 4.1). It is worth noting that not a single imprint of this cereal was found in P'iatnyts'ke I, the site with the least amount of rye in the PBS.

By contrast, barley demonstrates a certain stability. In three cases, barley represents about a half of the PBS, while in another three another it is just between a fifth and a quarter. In most other cases, barley is at about the same level as common wheat and rye, between one and two thirds of the PBS.

Barley and hulled wheat are the first cultivated plants grown on the territory of modern Ukraine.[31] This is mostly because of their extreme adaptability to all sorts of soil. Equally convenient is proso millet, which does not require deep plowing (in fact, deep plowing could considerably slow down the germination of millet grains).[32] Barley is quite resistant to climatic conditions, soil fertility and agrotechnical capabilities; it can grow on less fertile soils than those required by wheat, and it can withstand slight salinity. Because of that, barley is cultivated on large areas, and it is still one of the most important cereal crops in the world. Barley belongs to the drought-resistant and early maturing crops.[33] In southern areas, barley may sometimes be harvested twice a year. In Georgia, when wheat is not growing, barley is sown, even late in the year.[34] Moreover, barley may also be used for fodder, which means that it will preserve its significant role irrespective of agrotechnical capabilities.[35]

The cultivation of both naked, common wheat and rye indicates a more advanced level of soil cultivation, largely because of improved tools. Scholars have attributed the spread of rye cultivation to the introduction of plowshares that made plowing deeper and better.[36] In other words, a significant presence of rye and naked varieties of wheat in the PBS indicates an advanced form of agriculture. The cultivation of rye is also betrayed by finds of brome seeds. Both rye and field brome are the main weeds of winter crop fields, primarily of rye.[37] Archaeologists agree that any significant presence of brome is an indication of the cultivation of winter rye.[38] Rye can also be used as fodder, in the form of both grains and the flour from its straw.

Oats is also used as fodder. It is interesting to note that, taken together, barley, rye, and oats form between 50 and 70 percent of the PBS. The only exception is Chuhuiv, a site on which those cereals make up only 30 percent of the PBS. This may have something to do with the growth of the settlement inside the hillfort, which shifted the emphasis from production to consumption, with no need for fodder.[39]

31 Pashkevich 1992, 23.

32 Ielagin 1955, 9.

33 Kobylianskii and Luk'ianova 1990, 188.

34 Bregadze 1982, p. 81.

35 According to Dobiash-Rozhdestvenskaia 1936, 18, barley was used in the Middle Ages as fodder, because it could "feed cattle better than wheat." The nutritive properties of straw and chaff of barley are very close to those of hay (Vavilov 1986, 124).

36 Lange 1975; Iazhdzhevskii 1988, 9899.

37 Smirnov and Sosnikhina 1984, 5–7.

38 Kir'ianov 1959, 333; Kir'ianov 1967, 147; Mykhaylyna, Pashkevich, and Pyvovarov 2007, 60.

39 Svistun and Gorbanenko 2011.

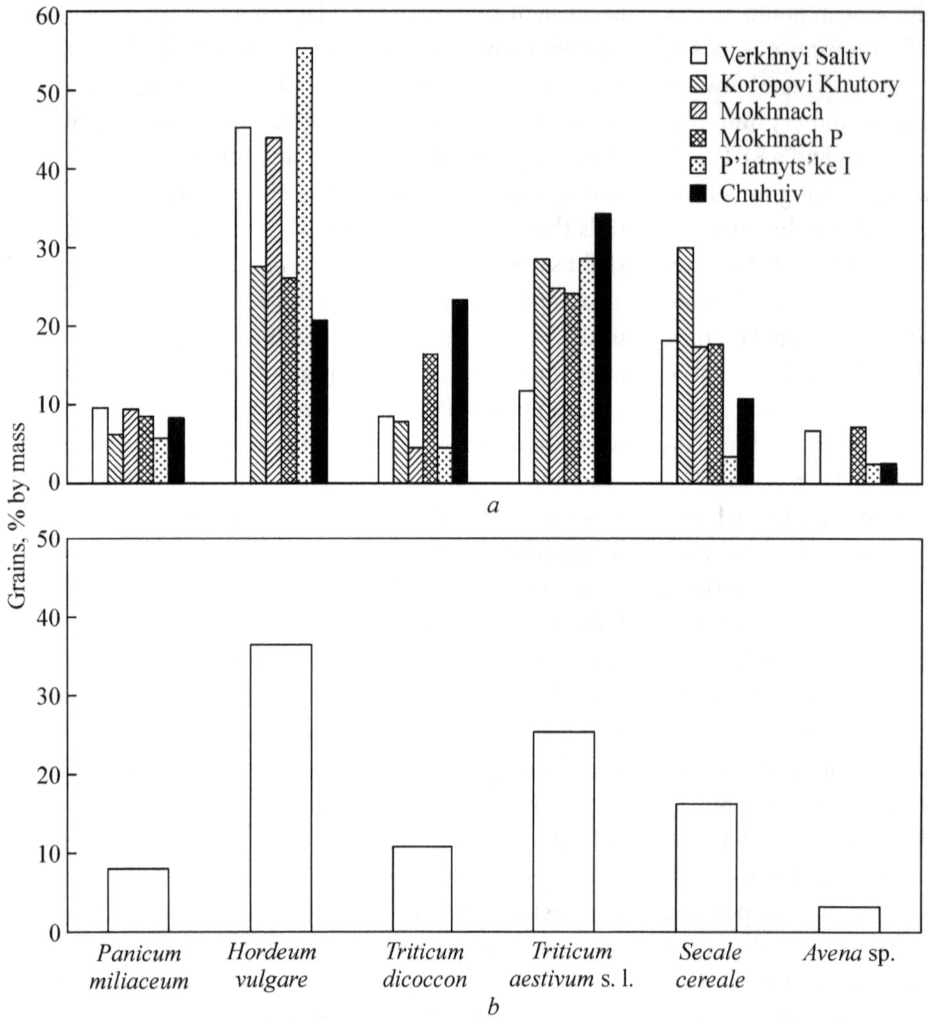

DIAGRAM 4.2 Palaeobotanical spectra of cereals from Saltiv sites in the region under study, by weight

4 Tools for Harvesting

Sickles and scythes are the tools that can be associated with harvesting. They have different forms and types, some common across Eastern Europe, others specific for Saltiv culture.

No less than 52 sickles are known for the area under study (Plate 4.6; 4.7).

Slightly less than half that number (21) have been found in Mokhnach P, 15 in Koropovi Khutory, 12 in the hillfort at Mokhnach, 3 in Verkhnyi Saltiv, and just one in P'iatnyts'ke I. Because of the varying state of preservation, only 47 sickles could be analyzed for typology, primarily on the basis of how the blade is attached to the handle (Table 4.2).

Most specimens are stemmed sickles (68 percent of all those specimens that can be assigned to one type or another). Almost a quarter (23 percent) of all sickles are of the shanked type, and only about 6 percent are toggle-action sickles.[40] Those proportions are about the same within the Saltiv culture as a whole (hooked sickles are extremely rare, in fact a relic for the area under study, and for the Saltiv culture in general).

40 In Eastern Europe, toggle-action sickles known from Saltiv sites. They are relatively common in burial assemblages with weapons and horse-riding equipment. That is why some interpreted them as typical for warrior graves (Shramko 1983; Mikheev 1985, 47–48). However, such sickles were found together with both horse gear and weapons, as well as with only weapons or only horse gear harness (e.g., in graves 151, 214, 252 and in features XII and XVI in Sukha Gomil'sha, for which see Aksenov and Mikheev 2006, 96). In other words, not all burials with such sickles are warrior graves. In our opinion, the appearance of toggle-action sickles in burials has a different explanation. First, the deposition of a folding sickle in the grave with weapons may be a symbol of land ownership. Second, its deposition in a grave with horse gear may have been meant to suggest its use for horse feed.

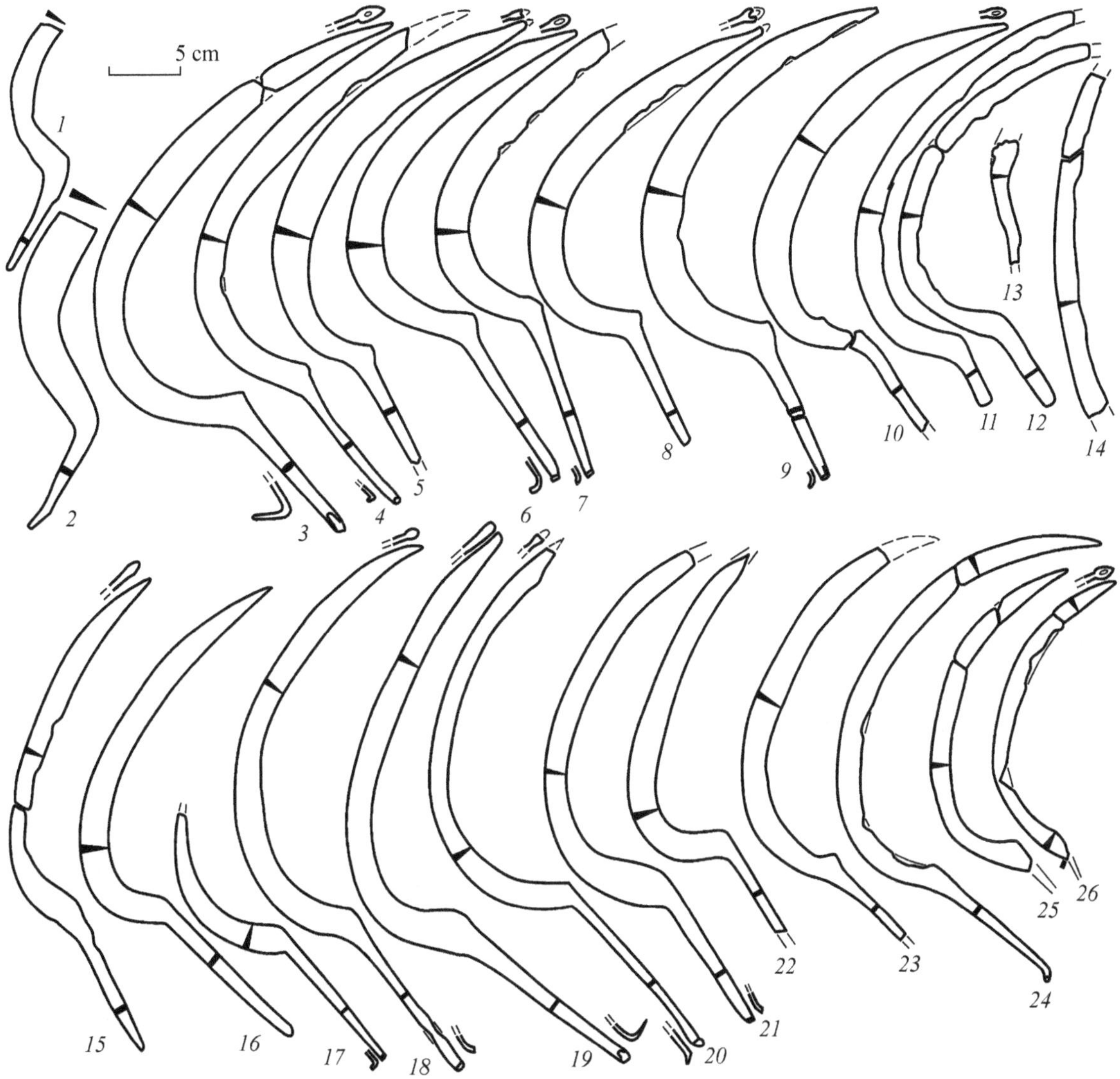

PLATE 4.6 Stemmed sickles: *1, 2* — Verkhnyi Saltiv; *3–10* — Koropovi Khutory; *11–16* — Mokhnach; *17–26* — Mokhnach P

TABLE 4.2 Sickles, according to the type of attachment

Site	Type					Total
	Stemmed	Shanked	Toggle-action	Hooked	Not defined	
Mokhnach P	12	7	—	1*	1	21
Koropovi Khutory	13	—	2	—	—	15
Mokhnach	5	4	—	—	3	12
Verkhnyi Saltiv	2	—	—	—	1	3
P'iatnyts'ke I	—	—	1	—	—	1**
Total	32	11	3	1	5	52

* — blank; ** — some more sickles of various condition of preservation were found in the cremation cemetery next to the settlement of Kochetok (Svistun 2012c).

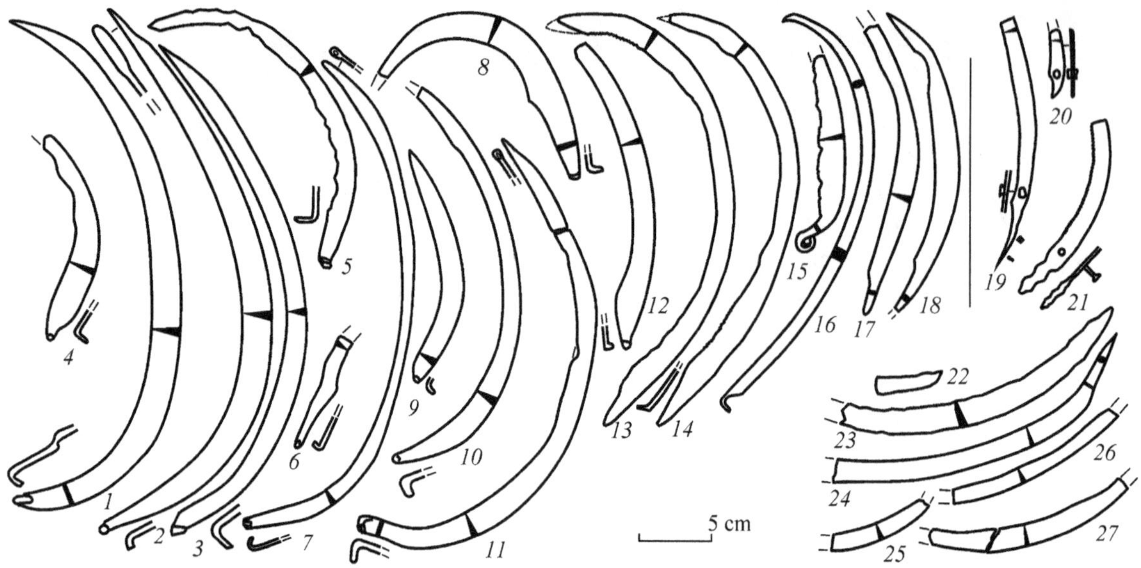

PLATE 4.7 Sickles with different types of attachment: *1–14* — shanked type; *15, 16* — hooked type; *17, 18* — primitive stemmed type (a handle continues the line of the blade); *19–21* — folding (toggle-action) type; *22–27* — unspecified fragments; *1–4, 24, 25* — Mokhnach; *5–11, 16–18, 26* — Mokhnach P (*16* — blank); *12, 27* — Mokhnach U; *13, 14* — Sukha Gomil'sha (accidental find); *15* — Shevchenkove 4; *19, 20* — Koropovi Khutory; *21* — P'iatnyts'ke I (grave); *22* — Verkhnyi Saltiv; *23* — Kochetok

TABLE 4.3 Scythes, according to the type of attachment

Site	Type				
	With heel		Without heel	Not defined	Total
	Stem	With spike			
Mokhnach	4	—	5	—	9
Mokhnach P	—	2	4	1	7
Verkhnyi Saltiv	1	1	—	—	2
Kochetok	1	—	—	—	1
Total	6	3	9	1	19

The shape and proportions of sickles measured or reconstructed graphically using analogies and visual observation, indicate the use of advanced forms, known both from the latest materials of Kievan Rus',[41] and from ethnographic materials, including those of the modern age.

The most common (stemmed) variety (Plate 4.6) appears practically on every site. As a matter of fac, any set or hoard of agricultural implements that includes sickles has at least one specimen of the stemmed type.[42] The hoard found in Koropovi Khutory has only stemmed sickles (13 specimens), all of the same size.[43] Blades of stemmed sickles are between 20 and 25 cm long, shorter blades being exceptional (a fixed minimal length of the working part is 11 cm (plate 4.6, *17*); probably, the blade was somewhat broken). The stem is between 7 and 11 cm long. On some specimens, the stem was longer than handle, which is between 9 and 10 cm long (Plate 4.6, *3, 4, 6, 7, 9, 17–20, 24*). Most stemmed sickles have a high arc of the blade, displaced towards the handle (i.e., the most productive form of blade, see Plate 4.6, *1–12, 15, 16, 18–26*). A few, however, have a barely curved blade, or the highest point in the middle of the arc, both rendering the tool inefficient (Plate 4.6, *17, 18*). The most advanced form of this tool is that with a higher blade arc, but there are specimens with a low arc (plate 4.7, *17, 18*). Those sickles were hardly efficient tools, and may be relicts of a much older harvesting technology. Some stemmed sickles have a loop at the end of the blade, the meaning (and purpose) of which remains unclear (Plate 4.6, *3, 5, 8, 10, 26*). Stemmed sickles were very common in Khazaria.[44] Such sickles were also used by the Slavs in Left-Bank Ukraine in the early Middle Ages.[45] Isolated finds of stemmed sickles are also known among from assemblages in the Don area that have been attributed to the Slavs.[46]

41 E.g., Levashova 1956, 60.
42 Koloda 2015.
43 Koloda 2014b, 42 and 44–45.
44 Mikheev 1985, 45 fig. 25, *8–16*.
45 Gorbanenko 2007, 46 figs. 25 and 26.
46 Vinnikov 1995, p. 40; Gorbanenko 2014c, fig. 3.

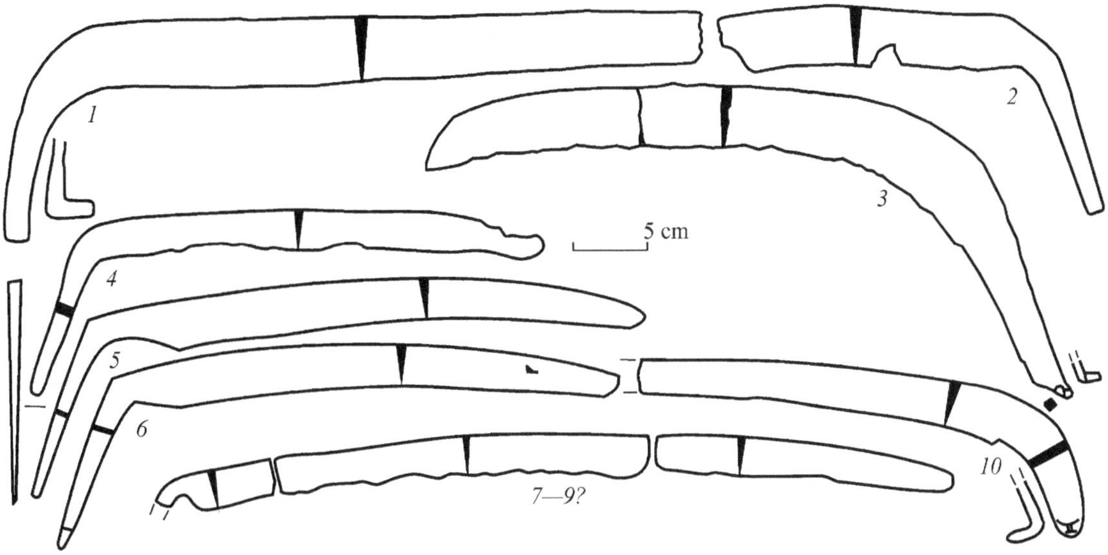

PLATE 4.8 Scythes with heel: *1, 2* — Verkhnyi Saltiv; *3* — Kochetok; *4–9* — Mokhnach; *10* — Mokhnach P

Shanked sickles have a less advanced shape — a low blade arc, with the highest point towards its center or, in one case, even to the end of the blade (Plate 4.7.*1–11*).[47] The blades of those sickles are longer than those of stemmed specimens — between 14 and 34 cm. It is possible that the increased blade length was supposed to compensate for the lower efficiency, when compared with stemmed sickles. An interesting detail is that like some stemmed, some shanked sickles have loops at the end of their blades (Plate 4.7, *7, 11*), others a flattened enlargement in a perpendicular plane (Plate 4.7, *2*). Shanked sickles are relatively common in the Saltiv culture.[48] They were deposited in cremation burials in Sukha Gomil'sha,[49] but are also known from the settlement in Mokhnach U (Plate 4.7, *12–14*). Outside Khazaria, in assemblages attributed to the Slavs, such sickles are rare.[50]

Toggle-action sickles have shorter blades, between 11 and 18 cm long.[51] Specimens known from the region under study are few and fragmented (Plate 4.7, *19–21*). Only one hooked sickle is known from Mokhnach P, and seems to have been a half-manufactured artifact (Plate 4.7, *16*).[52] The blade is about 26 cm long, but the blade arc has the highest point towards the outer end, which made this a rather inefficient tool. During the excavation of a kiln in Shevchenkove 4, another hooked sickle was found (Plate 4.7, *15*). This substantiates our idea, namely that the inhabitants of the Khaganate used antiquated reaping tools of archaic forms along with very advanced ones.

Scythes were found on three sites in the region, as well as among materials of the Kochetok cemetery, which is near P'iatnyts'ke I.[53] Volodymyr Mikheev has distinguished two types of scythes found on Saltiv sites — scythes with curved heel, and scythes without heel, but with a hole at the junction been blade and snath. He further distinguished variants for each type, on the basis of the shape of working part (straight or curved) and other details.[54] The accumulation of a great number of scythe finds from Khazaria since the publication of Mikheev's book invites a fresh look at his typology. To be sure, our type 1 includes scythes with heel (Plate 4.8), while type 2 that includes scythes without heel (Plate 4.9; see Plate 2.10, *2*). We have been able to distinguish two variants for the first type. The first is characterized by a heel completed with a shank hammered into the end of the snath (or fastened with an iron ring, as with modern scythes, or with a flexible strap: Plate 4.8, *2, 4–6*; see Plate 2.10, *1*). The other variant has a shank at the end of the heel, which is perpendicular to the main plane of the metal strip (Plate 4.8, *1, 3, 10*). After hammering the shank to the snath, the joint was secured with a belt (see Plate 2.10, *1*) or fastened with a ring (see Plate 2.11, *1*). Working parts of scythes of type 2 were attached to the snathe with a metal rivet and with additional

47 Such sickles are known in the region under study since the early Iron Age.
48 Mikheev 1985, 45 fig. 25, *2–7*.
49 Aksenov and Mikheeev 2006, p. 96; Koloda 2012, 31 fig. 3, *2, 3*.
50 Gorbanenko 2007, fig. 25, *1*; 26, *37*.
51 Mikheiev 1985, fig. 26.
52 Hooked sickles are the oldest such tools in Eastern Europe, the earliest specimens of which appear in the early Bronze Age.
53 Svistun 2012, fig. 6, *40*.
54 Mikheev 1985, 29. Only the types have passed the test of time, as most subsequent researchers employed Mikheev's typology, but without his variants.

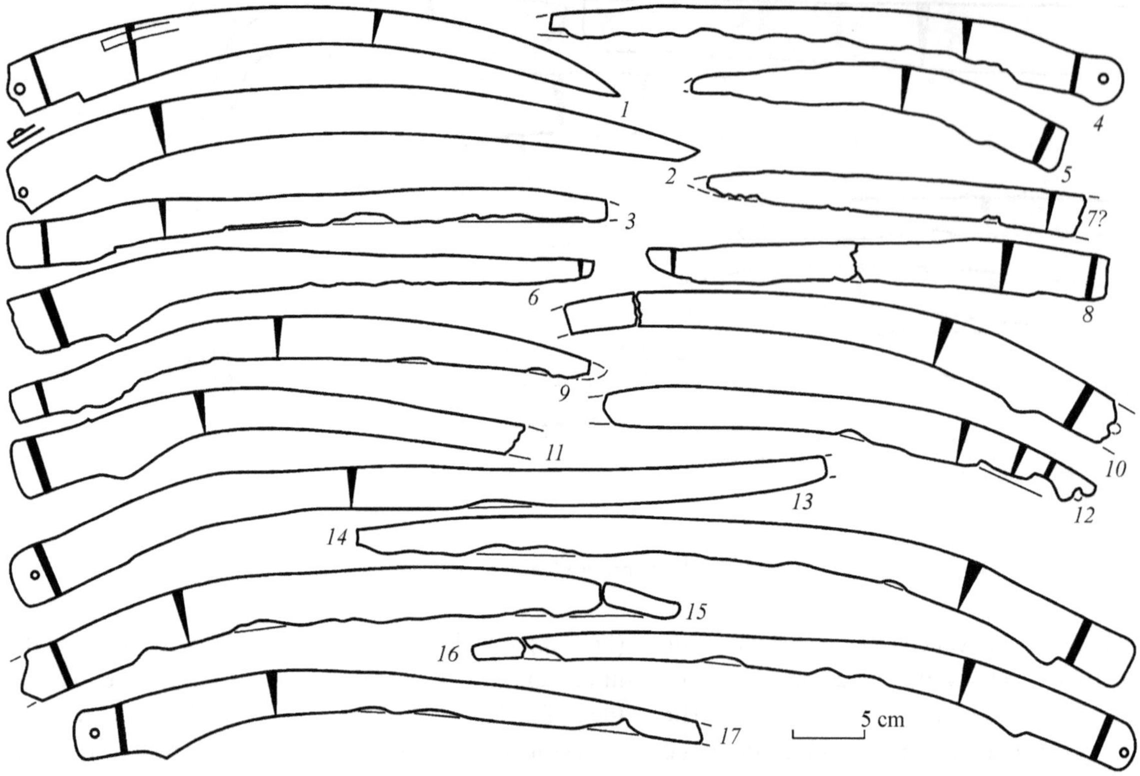

PLATE 4.9 Scythes without heel: *1–5* — Mokhnach; *6–17* — Mokhnach P

hard fastenings (Plate 2.10, *2*; Plate 2.11, *2*).[55] A specimen broken into three parts may be attributed to type 1 but not to any of its variants (plate 4.8, *7–9*).

Scythes of the two types are equally efficient, a quality that depended, to a great degree, on the blade length. Scythes of type 1 have between 26 and 51 cm-long blades, and those of type 2 between 21 and 55 cm-long blades. The width of the blade varies between 1.5 and 4 cm, but one should take into account the wear resulting from the use of tool. The scythe from Kochetok is a certain exception, with its 5 cm-wide blade (Plate 4.8, *3*). Scythes such as those analyzed above have been found on many Saltiv sites.[56] Plates 2.10 and 2.11 show the graphic reconstruction of the tools and the ways they were fastened. Leaving aside the differences between types, all scythes are in fact *gorbushy*. The blade and snath are on the same plane, which forces the user to bend over, when handling the *gorbusha*. Both types of those *gorbushy* existed until the first ethnographic reports of modern times.[57] In fact, such shapes are still in use nowadays. To judge from the ethnographic data, both varieties of the scythes (with or without special devices in the form of small rakes) could be used for harvesting.[58] However, more often than not, scythes were used for the production of hay to feed stabled animals during winter.

5 Crop Storing

After harvest, crops were stored in various ways. Silos, barns and, probably, large cellars were used for large amounts of grain. For smaller (individual or family) stocks or for quick use, so-called inner cellars and pits inside dwellings and outbuildings were used, in addition to various containers. Some of those containers were ceramic (mostly large earthenware jars), others were made of soft materials (wicker boxes or textile and leather bags and sacks).

Silos were the best solution for long-term storage. They can be recognized archaeologically by the large volume, as well as the round plan (1.5 to 2.0 m in diameter at the opening), depth (1.5 to 2.3 m),[59] and a flat (sometimes con-

55 Mikheev 1985, 29–31.

56 Mikheev 1985, fig. 21; Pletneva 1994, fig. 47, *5–7*; Flerov 1994, fig. 13, *1*.

57 Zelenin 1991, p. 62, fig. 12; 13.

58 Zelenin 1991, p. 62.

59 The shallowest grain pit is 1.30 m deep.

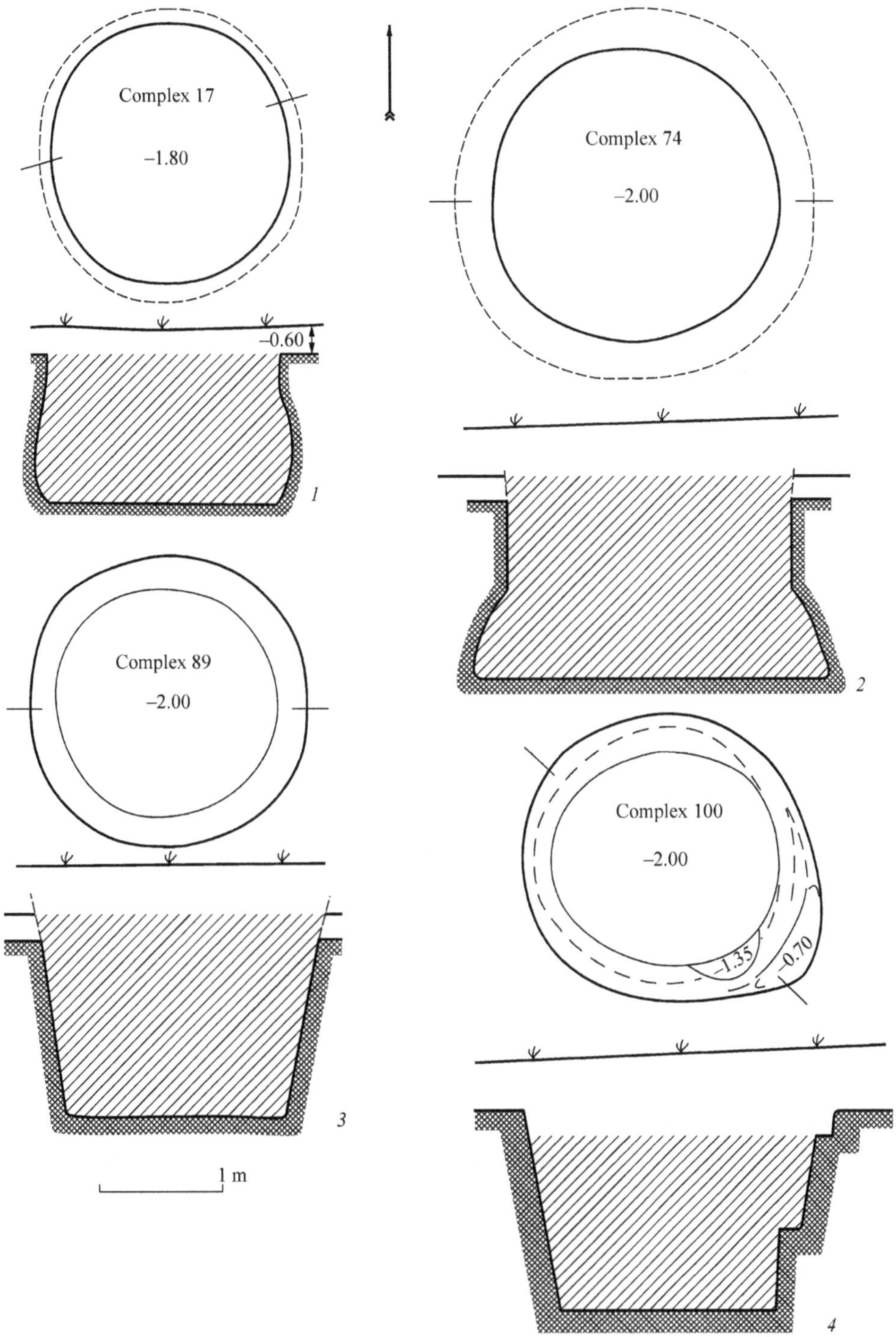

FIGURE 4.1
Mokhnach, silos of types I (*1, 2*) and II (*3* — II.1; *4* — II.2)

cave) bottom. The filling of a silo is typically clear, as long as the silo was not turned into a refuse pit. However, some pits contained fragments of millstones that were probably deposited there symbolically (Fig. 4.1). Silos come in different shapes. Some are cylindrical, with a recess in one (Fig. 4.1, *1*) or several sides (Fig. 4.1, *2*). The section of others looks like a truncated, inverted cone. Of those, some have smooth walls (Fig. 4.1, *3*), while others have steps on one or several sides (Fig. 4.1, *4*). There are also silos with round section (Fig. 4.1, *2*) or in the form of a truncated cone (fig. 4.2, *1*). Similar silos are known from Dmitrievka and Mayaki.[60] They have also been found on Saltiv settlements in the steppe zone and in the Crimea (mostly at Bakla).[61]

60 Pletneva 1989, 46–48; Vinnikov and Pletneva 1998, 121 and 129.
61 Krasil'nikova 2005, 12; Guskov 2007, 23 and 29–33.

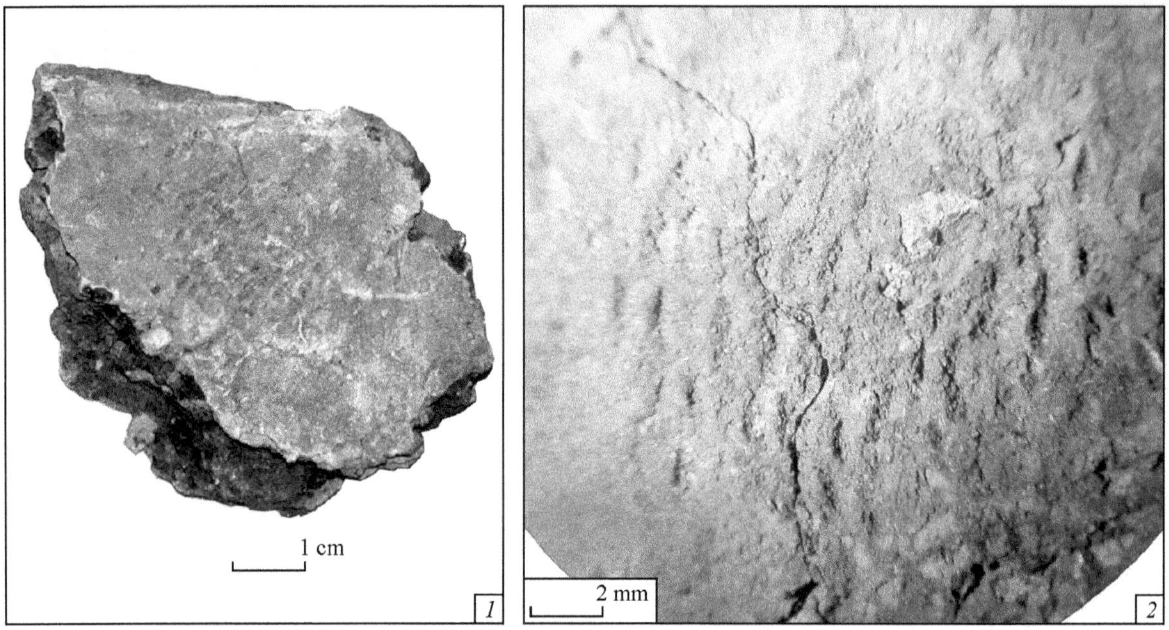

PHOTO 4.1 Fragment of a jar with a burlap imprint: (left) general view; (right) enlarged print of the fabric imprint
(LEFT) PHOTO BY V. V. KOLODA; (RIGHT) PHOTO BY T. M. KRUPA

The archaeological signature of barns is the presence in each one of them of two grain pits, which were most likely used for simultaneous storage. So far, three such features are known (Fig. 4.3; 4.4). The depth of the pits in those barns varies between 1.8 and 2.1 m, which brings them close to silos.

A number of dwellings have been found on several Saltiv settlement sites that have outside cellars. The largest of them may have been used to store grain. Those cellars are different from both silos and barns, as their sections are typically asymmetrical and their sides have steps that may have served as shelves for food storage in containers (Fig. 4.5; 4.6). Moreover, cellars are much deeper than silos (1.7 to 3.4 m).

Inner cellars and pits for the storage of grain appear occasionally inside could dwellings (Fig. 4.7; 4.8) or outbuildings (Fig. 4.9; 4.10). They are typically smaller than any other storage facilities, most likely because of their specific location. However, their sections are not unlike those of silos.

To store a relatively small amount of grain for everyday use, an earthenware jar may have been used, or large kitchenware pots with combed ornamentation (Plate 4.10). Neither boxes made of withes and bast, nor textile or leather bags or left any archaeological traces, but the existence of rough fabric for sacking (burlap) is attested by an imprint on the wall of an earthenware jar from feature 20 in the Mokhnach hillfort (the largest such jar in that feature; Photo 4.1).[62] Moreover, there is abundant ethnographic evidence of agricultural products being stored in containers of organic materials, such as bags or sacks.

The many ways of storing grain discussed above are illustrated in the archaeological record of the region under study. All have been documented, for example, during the excavations in the Mokhnach hillfort (Map 1.7). In the southwestern part of the area enclosed by ramparts, large silos, barns and deep cellars have been found. Up to seven storage facilities have been documented for one dwelling and six separate features have been found in one area, all next to each other and used for storing grain. In the northern part, there were only two silos, but many more thick-walled, earthenware jars have been found (mainly in dwellings).[63]

62 According to Tat'iana M. Krupa, the coarse canvas belongs to O. I. Davidan's fourth grade, and may thus be defined as burlap. The observed fibers were undoubtedly of vegetable origin, perhaps hemp. We wish to express our gratitude to Tat'iana Krupa for her analysis of the imprint. Burlap imprints have also been documented in the Sverdlovka 1 hillfort (Chernihiv region, Ukraine; Chernenko et al. 2017, fig. 16) There is also indirect evidence of sacking on contemporaneous Slavic sites. On the floor of out building 1 in the Novotroits'ke hillfort (Sumy region, Ukraine) was a pile of rye grain with a hoe-head on the top (Liapushkin 1958c, 150). This is most likely an indication that the grain was in a sack.

63 Koloda and Gorbanenko 2010, fig. 50.

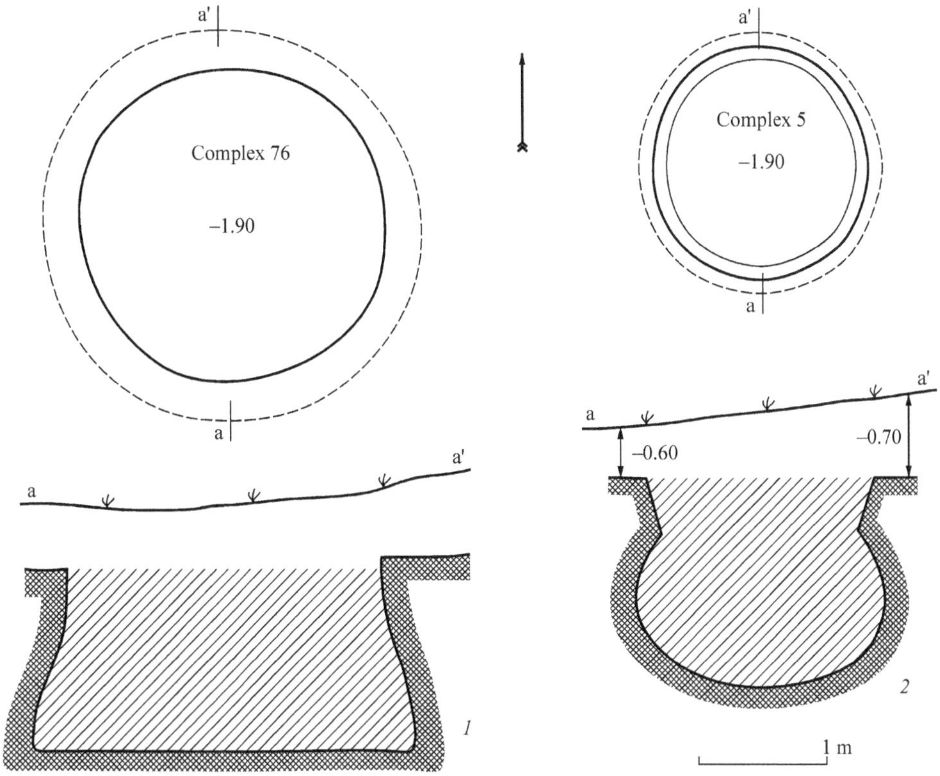

FIGURE 4.2
Mokhnach, silos of types III (*1*) and IV (*2*)

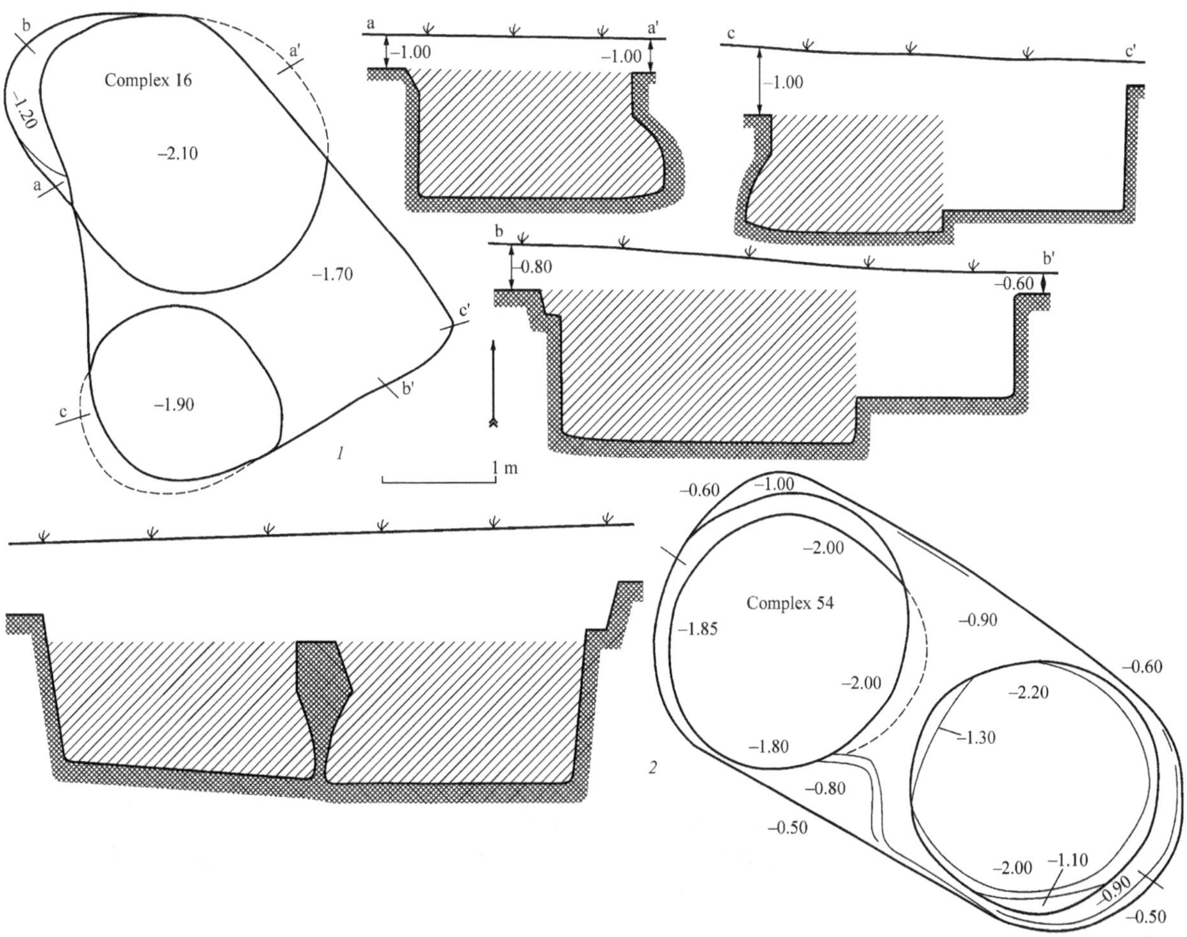

FIGURE 4.3 Mokhnach, barns (*1*, *2*)

TABLE 4.4 Capacity of household facilities for storing grain discovered in the northern yard of the Mokhnach hillfort

Number of feature or dig	Amount of fragments of millstones	Purpose	Minimum useful volume (in liters)	Average weight of stock (in kg)	Notes
5 / 10	—	Grain pit	3000	2310	Complex of household structures
7 / 10	3	The same	7500	5775	The same
10 / 10	—	Cellar	7500	5775	The same
12 / 10	—	Grain pit	5000	3850	Near dwelling (feature 13)
16 / 10	1	Barn with two inner pits	1600 + 4400	1232 + 3388	Complex of household structures
17 / 10	—	Grain pit	2800	2156	The same
23 / 10	1	The same	5000	3850	The same
28 / 10	1	Small cellar	714	550	Near dwelling (feature 25)
29 / 10	1	Grain pit	6100	4697	The same
39 / 10	—	The same	4180	3219	The same
41 / 10	—	The same	2094	1613	The same
43 / 10	—	The same	2638	2032	The same
53 / 10	—	The same	6384	4915	Complex of household structures near shaft
54 / 10	—	Barn with two inner pits	3647 + 4357	2808 + 3355	The same
65 / 10	—	Cellar	4933	3798	The same
67 / 10	3	Grain pit	7709	5936	The same
69 / 10	1	Barn with two inner pits	1857 + 281	1430 + 217	Near dwelling (complex 64)
72 / 10	—	Grain pit	2218	1708	Complex of household structures near dwelling (feature 79)
73 / 10	—	The same	2786	2145	The same
74 / 10	1	The same	3312	2550	The same
76 / 10	—	The same	2989	2302	The same
81 / 10	—	The same	2825	2175	The same
82 / 10	—	The same	1964	1512	The same
84 / 10	—	The same	1666	1283	The same
89 / 10	—	The same	2166	1668	
91 / 10	—	The same	1055	812	Complex of household pits in the south-east part of the dig
92 / 10	—	The same	1419	1093	The same
93 / 10	—	The same	1205	928	The same
100 / 10	—	The same	1901	1464	Near dwellings (features 94 and 98)
40 / 11	—	The same	1453	1119	Complex of household structures near dwelling (feature 27)
41 / 11	2	The same	2519	1940	The same

Silos, barns, and large cellars indicate the need for storage of large quantities of grain, and that is an indirect evidence of the specialization and commodification. By contrast, smaller storage facilities (for example, in ceramic containers) point to a preferential consumption. When plotted on the map of the hillfort, this polarity may also indicate the residential quarters of different groups of people, those who were not agriculturists (artisans, but also, perhaps, pastoralists), and soil cultivators, respectively.

Up to 2016, no less than 31 storage features of different forms and sizes were identified on the northern (Saltiv) yard of the Mokhnach hillfort.[64] They can be classified as follows (Table 4.4; Figs. 4.1–4.6): 80.6 percent are silos, and only about 9.7 percent are barns and cellars. This clearly

64 Some of them were taken into account in Koloda and Gorbanenko 2010, figs. 46–49; and Gorbanenko and Koloda 2013, figs. 60–63.

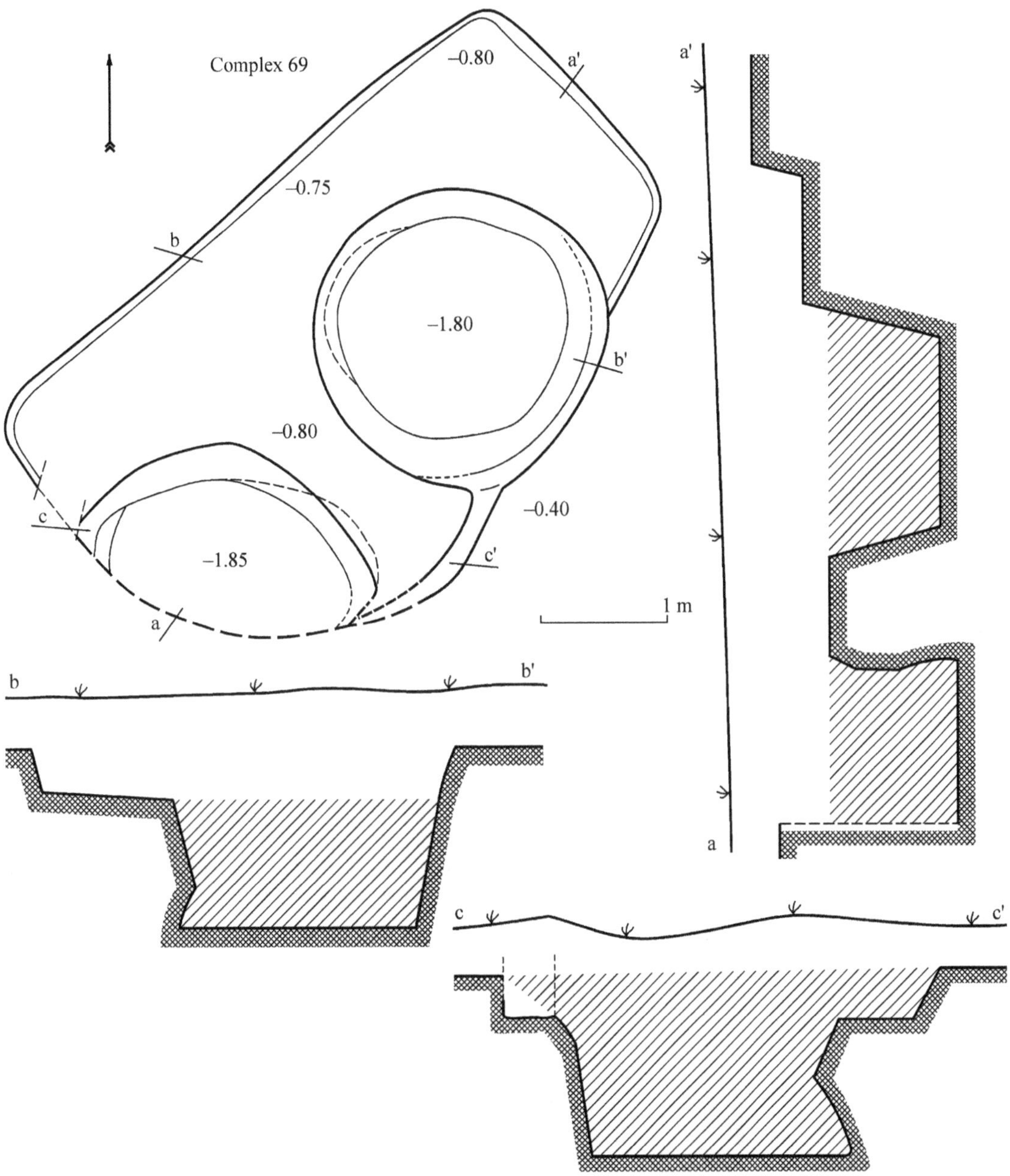

FIGURE 4.4
Mokhnach, barns

shows that silos were the preferred form to store grain in Mokhnach. The capacity of those silos varies between 1,055 to 7,500 liters, which could store from 812 to 5,775 kg of grain (Table 4.4).[65]

About 40 percent of all ceramic remains from the southern area of the excavation in Mokhnach (Map 1.7) are either earthenware jars or amphorae, with a ratio of 1 to 4.[66] In the northern, the ratio changes to 2 to 3, but the percentage of containers in the ceramic material also increases to about 55. It goes without saying that entire vessels are extremely rare, and were all found in the northern area: a great pot (Plate 4.10, *5*), next to a jar (Plate 4.10, *2*) inside the house-workshop of a blacksmith.[67] Using a simple calculation principle,[68] the capacity of the pot in question was 13 l, which allows one to store up to 10 kg of grain inside it.

The capacity of the jar was 61 l, the equivalent of 46 kg of corn. In 2007, in corner pits of one of the dwellings two other jars were found (plate 4.10, *1*, *3*). The smaller one had a capacity of 39 l, the larger one 42 l. The jars could store was 30 and 32 kg of grain, respectively.

65 To obtain this estimate, we relied on the idea that on average 77 kg of grain would fit in 100 l (Brockhaus and Efron 2002).

66 Earthenware jars are the main kind of container for storing of dry bulk material, such as grain.

67 Koloda 2002b.

68 The vessel capacity is calculated from bottom to neck. To simplify calculation, the internal space was divided into two truncated cones with the bases on the shoulders.

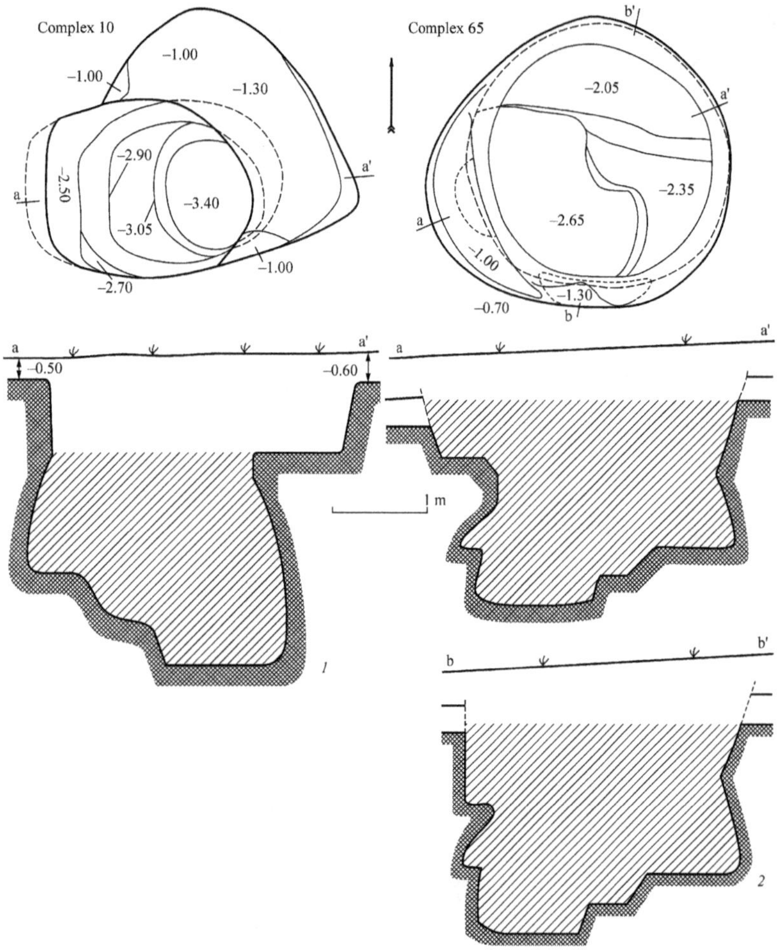

FIGURE 4.5
Mokhnach, cellars (*1, 2*)

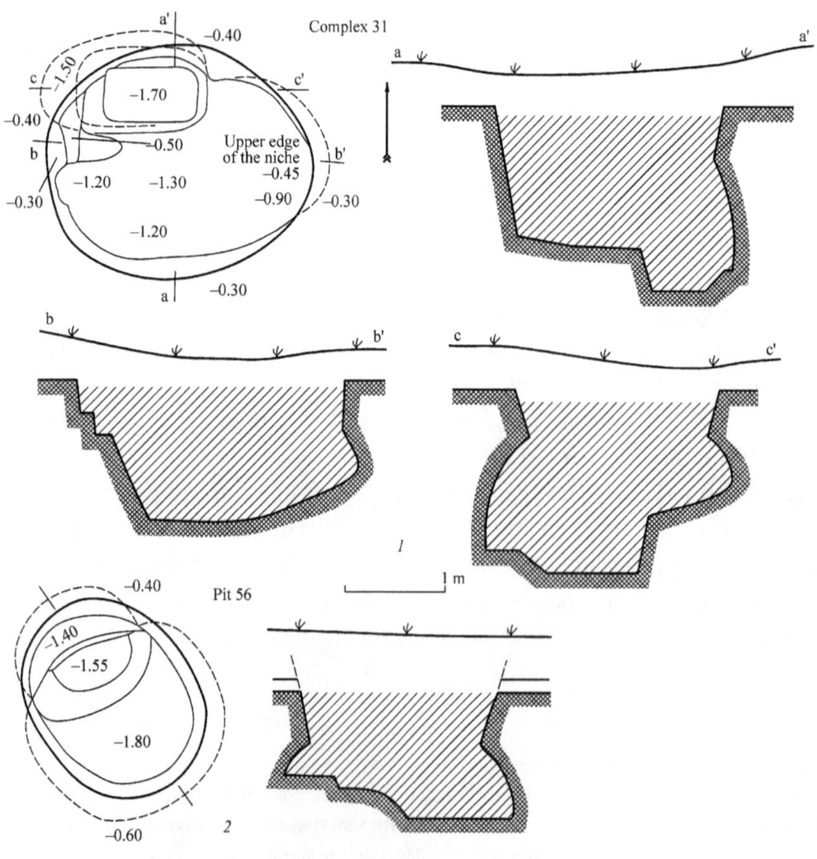

FIGURE 4.6
Cellars: *1* — Mokhnach P; *2* — P'iatnyts'ke I

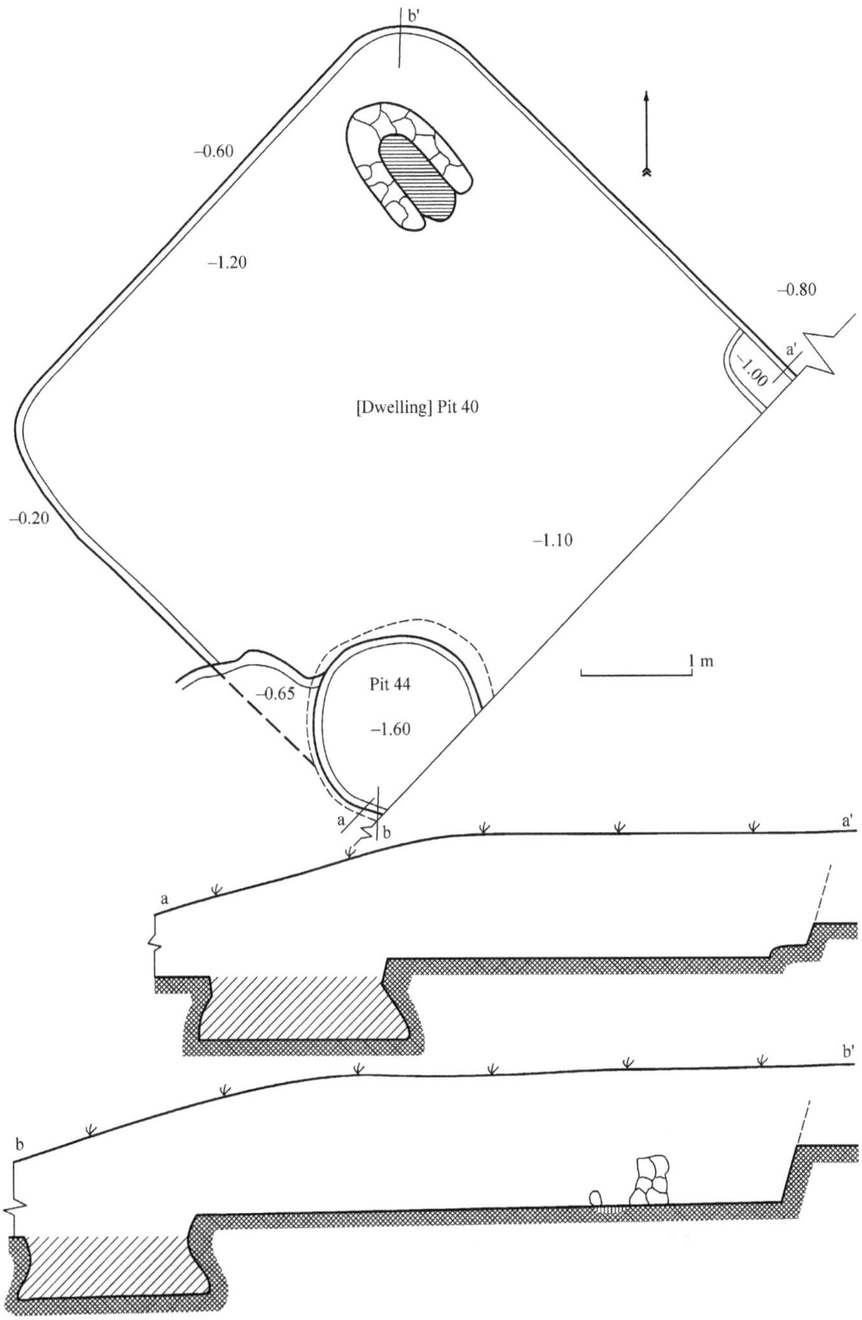

FIGURE 4.7
Grain pits in dwellings, P'iatnyts'ke I

Inside the hillfort, archaeologists also found storage facilities inside houses 47 and 101 (Fig. 4.9).

At Verkhnyi Saltiv both silos and containers have been found. Judging by A. T. Braychevska's report, there were numerous silos on the site.[69] In 1997 more than 500 fragments of thick-walled earthenware jars have been found within one of the early medieval yards, over an area of 4 sq m. After restoration in the lab, it became clear that those fragments belonged to at least seven jars. The feature in question may well have been a light, above-ground building for grain storage, which consisted of jars under a roof. As some of those jars could not be restored completely, it is impossible to calculate their capacity.[70] However, the 2004 excavation within the same area unearthed an intact jar (Plate 4.10, *4*). Its capacity was about 80 l, which means that more than 61 kg of grain could be stored inside it.

A utility room for storage of grain in soft containers may have also been found during the excavations of the late 1950s and early 1960s. One of the features revealed at that time was a sunken-floored building of elongated shape, with no fireplace. The excavator concluded that that was not a dwelling, but "some kind of outbuilding,

69 Koloda and Gorbanenko 2010, 72.

70 Gorbanenko and Koloda 2013, 91 fig. 58, *1–4*.

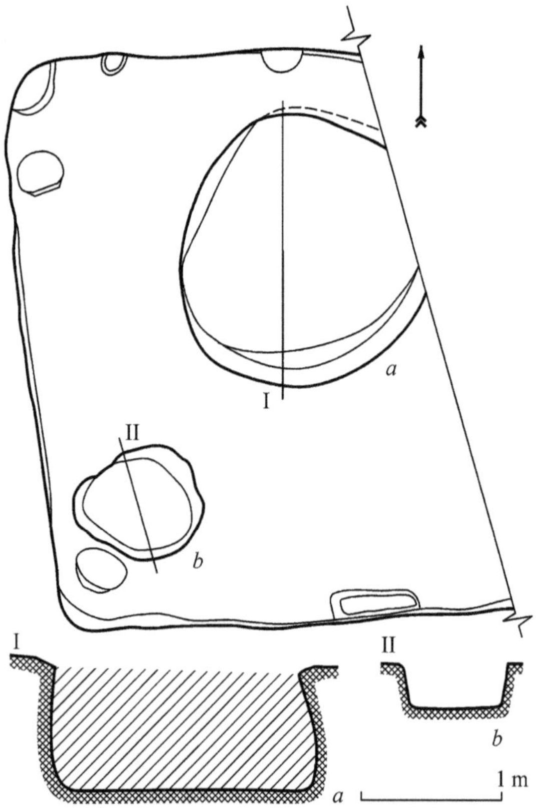

FIGURE 4.8 Grain pits in dwellings, Chuhuiv

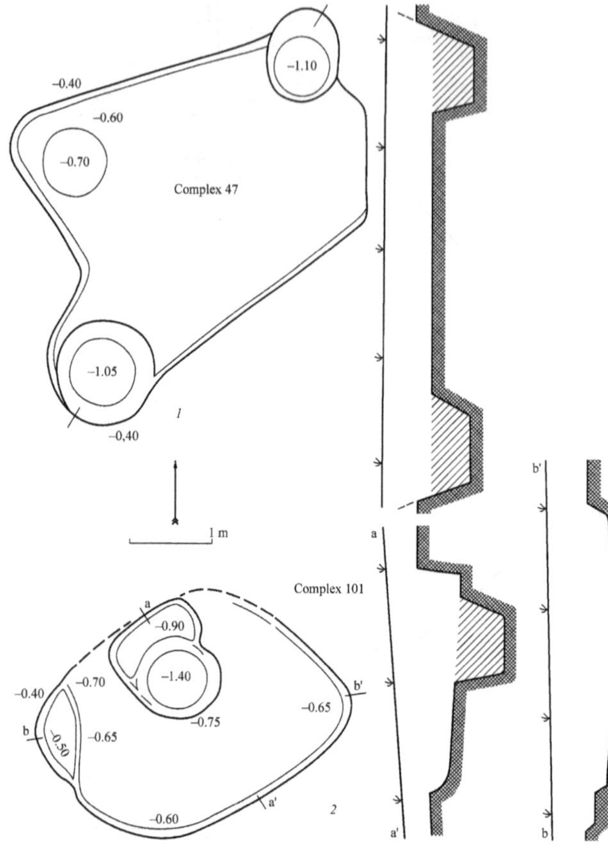

FIGURE 4.9 Grain pits in utility rooms, Mokhnach (*1, 2*)

consisting of several rooms; a type of shed or barn, where various property and bread were stored." A large amount of ash found inside the feature was believed to be from burnt un-milled grain, which was most likely stored in that feature.[71]

Silos, cellars, and storage buildings have been identified in P'iatnyts'ke I. As of 2016, no less than 10 silos and more than 10 cellars (?) have been found on the site.[72] In terms of types and capacity, the silos found in P'iatnyts'ke I are similar to those from the Mokhnach hillfort. Not all cellars on the settlement, however, can be confidently attributed to granaries. However, at least one of them (Fig. 4.6, *2*) may be interpreted as used for storing grain, judging by analogies from the Mokhnach hillfort (Fig. 4.5, *2*). For storage and everyday use, both indoor grain pits and jars were used. One jar was found almost intact (Plate 4.10, *6*). Its capacity is about 55 l, which would allow for the storage 42 kg of grain. It should be noted that, in general, fragments of jars account for a small percentage of ceramic remains found in P'iatnyts'ke I. However, there is a house with an indoor pit that was used for food storage (fig. 4.7). It looked very similar to a silo, but much smaller, with a capacity of only 563 l, which allowed for storing of 433 kg of grain inside it.

Grain storing in the Chuhuiv hillfort was very similar (fig. 4.8). A granary pit was found in the central part of a sunken-floored building with frame-and-pole construction.[73] The opening had small bevels, intended to accommodate a wooden lid at the level of the house floor. The pit had a round plan and a large volume (more than 2 cubic meters, the equivalent of to 20 hectoliters or 1,540 kg of grain). A depression was also found inside the pit, most likely intended to accommodate a jar, the remains of which were found nearby.

An example of a house with a pit for grain storing is also known from a settlement in the steppe zone, Rohalyk (Donetsk region, Ukraine). This granary, however, was outside the house.[74] Another house with a storage in the eastern corner was excavated in Iutanovka (Belgorod region, Russia).[75] The most difficult problems of interpretation

71 Koloda and Gorbanenko 2010, 72–73, fig. 38. The quote is from A. T. Braychevs'ka's excavation report.

72 Many thanks to the excavator of the site, Viktor I. Kvitkovs'kiy, for kindly offering an update on the excavation results. 90given information.

73 Svistun and Gorbanenko 2011, 77 fig. 3.

74 Krasil'nikov 1981, fig. 6, 2.

75 Vinnikov and Stepovoi 2012, p. 173, fig. 52.

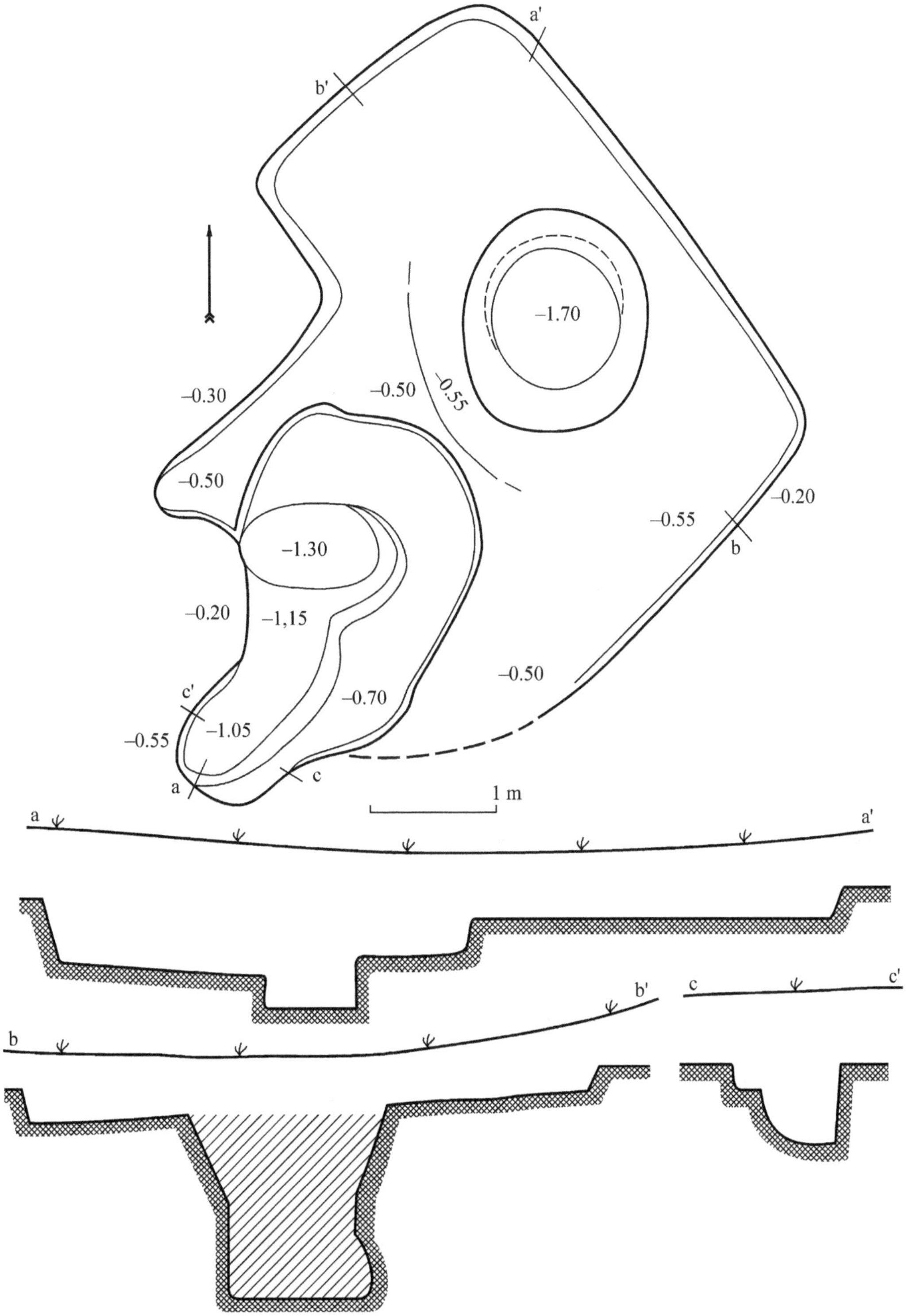

FIGURE 4.10 Grain pits in utility rooms, Mokhnach P

concern the Koropovi Khutory and Mokhnach P settlements. No deep underground features have been found, like those in the Mokhnach hillfort, in P'iatnyts'ke I or in Verkhnyi Saltiv. Perhaps the local, sandy and loamy soil was not conducive for digging deep granaries. However, consistently high percentages (30 to 40 percent) of jars have appeared in the ceramic material from all excavations in Koropovi Khutory. If one excludes the filling of small features (small pits, from which containers were naturally absent), the percentage reaches and even surpasses 50 (Table 4.5; 4.6). By contrast, ceramic containers are rare in Mokhnach P. Two features have been found

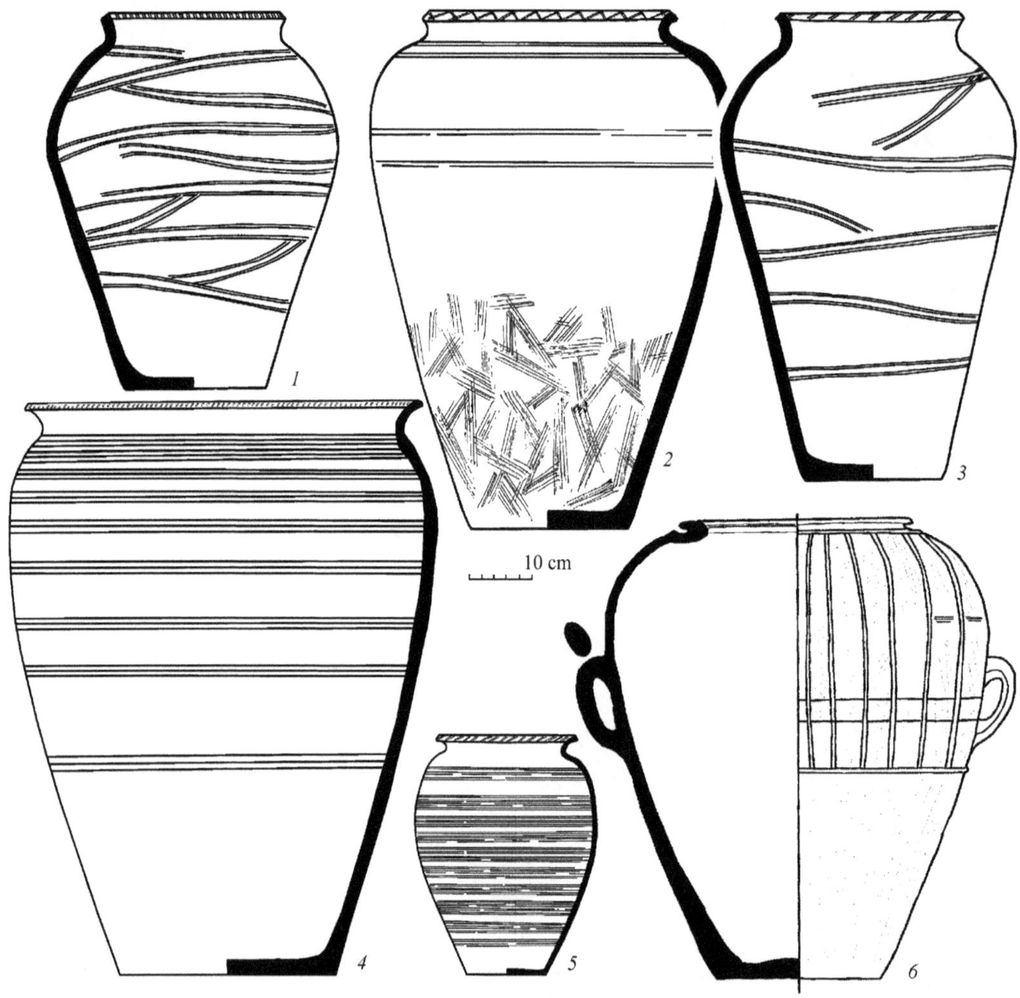

PLATE 4.10 Earthenware jars: *1–3*, *5* — Mokhnach; *4* — Verkhnyi Saltiv; *6* — P'iatnyts'ke I

TABLE 4.5 Fragments of the earthenware jars in the cultural layer of trenches 5 and 6 on the Koropovi Khutory settlement site

Year	Trench	Jars. %	Another artifacts of arable farming
2005	5	30	Fragment of millstone (no. 27. 37)*
2006	5	≈33	Fragment of sickle (no. 146). fragment of arable tool (no. 203)
2007	5	40	Grater (no. 233. 234). fragment of millstone (no. 283)
2007	6	37	—

Statistics obtained as a result of processing data from the description of material taken into account, but not included.
* Numbering after the field description.

there, in which grain was supposedly stored. One of them (feature 28) had two deep pits inside, at least one of which may have been used for storing grain (Fig. 4.10). A deep cellar (feature 31; Fig. 4.6, *1*) may also be interpreted as granary. Given the obvious craft activities performed inside the settlement,[76] the two features were most likely used for storing small quantities of grain, most likely in soft containers.

We have already noted that a choice for specific storage facilities may indicate the level of social development and professional orientation, but this idea needs further, independent investigation.

6 Crop Processing

The inhabitants of Saltiv settlements used graters and millstones to grind the grain, with a clearly stronger

76 Koloda 2010.

TABLE 4.6 Fragments of earthenware jars in features of trench 5 on the Koropovi Khutory settlement site

Number of complex	Purpose	Jars %	Other artifacts
1	Household pit alongside with dwelling	≈50	—
2	Dwelling	≈30. 50	Millstone (No. 140)*
3. 13**	»	≈30. 40	—
7	Household pit	>50	—
10	Cellar	≈26	Millstone (No. 195)
12	Dwelling	≈33	—
14	Outbuilding	42	—
17	Summer kitchen	>50	—
18	The same	45	—
19	Dwelling	42	—
20	Summer kitchen	≈43	—
22	Outbuilding	≈32	—
23	Dwelling	43	—
24	Outbuilding (cellar)	≈40	—
25	Outbuilding	≈33	—

Small household pits, where separate fragments of jars have been found, were not taken into account.
* Numbering after the field description; ** a complex was excavated for 2 seasons, so it got 2 numbers; percentages are given separately.

emphasis on the latter (Plate 4.11). The quern stones were of quartzite, with a few made of fine-grained sandstone, limestone and granite. Judging by the upper stone, those querns belong to the groups I and III, with appropriate reconstruction (see Fig. 2.13).

Graters are rare on the Saltiv sites, including those in the region under study (Plate 4.12; see reconstruction in Fig. 2.12). Quern stones in different manufacturing stages have been found inside a house in Verkhnyi Saltiv. According to the excavation documentation, in addition to quern stones, this feature produced large pottery containers.[77] In Koropovi Khutory, the fragment of a millstone was found in an outbuilding, where processing of grain may have taken place, as suggested by fragments of earthenware jars found in the outbuilding.[78]

Archaeologists have paid limited attention to the reconstruction of the crop processing. An outbuilding with millstones has been reconstructed, which has been attributed to the Zarubintsy culture (1st century BC to 3rd century AD).[79] Another outbuilding was reconstructed on the basis of materials attributed to the Sântana de Mureş-Chernyakhov culture.[80] The situation in the Pastyrs'ke hillfort, where quern stones were found in sunken-floored building, are chronologically closer to period of interest in this book.[81] All those examples have in common the fact that crop processing took place indoors next to storage facilities, such as pits or jars.

Mokhnach P occupies a special position in a discussion of crop processing. In addition to many fragments of quern stones, the excavations on the site produced a significant amount of waste resulting from their manufacturing process — pieces of broken stones (sandstone, quartzite, and occasionally granite). In fact, the latter is the most common category of finds on this site (51 to 52 percent of all finds), a situation without any parallel on any known Saltiv site. It may be concluded that Mokhnach P specialized in the production of millstones for the communities in the agricultural district around the hillfort of Mokhnach hillfort.[82]

A number of general conclusions regarding arable farming may be drawn at this point. First the advanced agricultural technology allowed for deep plowing, which was required for the cultivation of productive, but sensitive crops such as rye and naked varieties of wheat. Second, the reconstructed plowing tools made possible the turning the turf either complete or, at least, partially,

77 Gorbanenko and Koloda 2013, 98.
78 Koloda and Gorbanenko 2012, 128–129, fig. 12.
79 Kozak and Pashkevich 1985; Kozak 2008, 93–94, fig. 32.
80 Symonovych 1952.
81 Gorbanenko 2008.
82 Koloda 2010a.

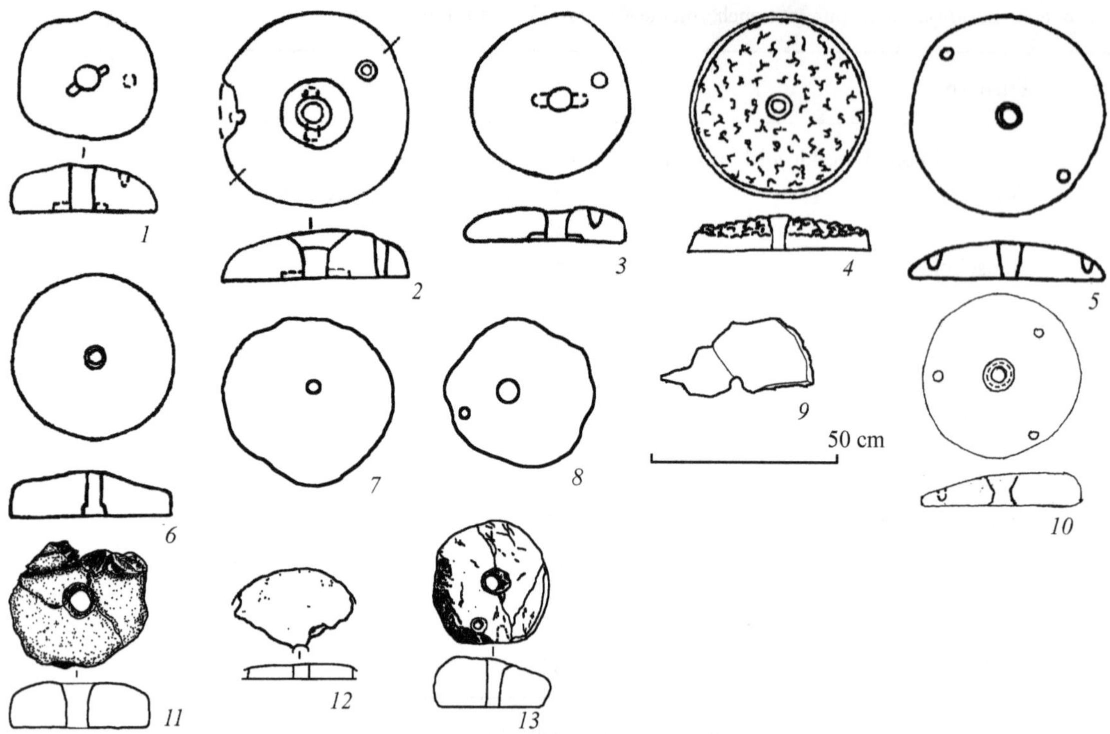

PLATE 4.11 Millstones: *1–6* — Verkhnyi Saltiv; *7, 8* — P'iatnyts'ke I; *9, 10* — Mokhnach; *11–13* — Chuhuiv

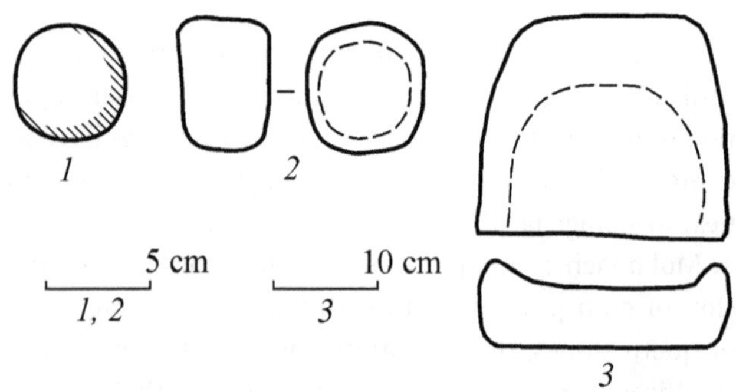

PLATE 4.12 Graters: *1* — P'iatnyts'ke I; *2* — Koropovi Khutory (graters); *3* — Verkhnyi Saltiv (plate)

thus contributing to the higher quality of plowing, and partial destruction of the root system of the weeds. Third, land plots were used continuously for longer periods, as indicated by imprints of weed seeds on ceramic artifacts. Moreover, winter weeds identified by such means suggest the introduction of the two- and three-field rotation. It is possibly that in both kinds of rotation, one part was left fallow for the partial restoration of fertility, although manuring may not be excluded (see below). The implication of the existence of spring and winter crops is crop rotation, which indicates a subtle understanding of the natural process of maintaining the fertility of the soil. Fourth, harvesting tools (sickles, and probably scythes) were different from other areas, and quite efficient, which ensured a significant productivity. Fifth, crop storage highlights a certain degree of economic specialization. Finally, grain processing was based primarily on rotary millstones, the manufacture of which is another indication of economic specialization.

CHAPTER 5

Animal Husbandry

A great deal has changed in the study of the animal husbandry in Khazaria because of the remarkable development of zooarchaeology over the last few decades. To be sure, the analysis of animal bones was traditionally done by specialists, who authored reports commonly published at the end of site monographs. For example, Valentyna Bibikova was in charge with the analysis of animal bones found during the salvage excavations on the site of the future Pechenega reservoir carried between 1959 and 1961 by the expedition headed by D. T. Berezovets (in fact, this was the study of economic zone of the Verkhnyi Saltiv early medieval complex). Her zooarchaeological identifications were then incorporated into Berezovets's field reports. Similarly, Roman V. Kroitor identified the bones from assemblages discovered in the Donets region.[1]

In a few cases, the analysis of animal bones became the basis of stand-alone zooarchaeological studies.[2] One of them, although published only as a summary, even attempted to compare finds from three sites — the campsite in Lomakino (Rostov region, Russia), the hillfort in Nizhnegnilovskoe hillfort (on the outskirts of Rostov-on-the-Don, Russia), and the Tsimliansk hillfort (30 km from Volgodonsk, Russia).[3] In addition to the interpretation of the material, valuable statistics regarding the bone assemblages from all three sites are given. To this day, all published assemblages of faunal material comes from Saltiv sites in Don region of present-day Russia; only Verkhnyi Saltiv is in modern Ukraine (and in the forest-steppe belt). A large study recently published on the zooarchaeological data from Saltiv sites in forest-steppe along the Donets is therefore unique so far.[4]

To make things worse, a number of zooarchaeologists have recently abandoned the idea of "minimum number of animals" (MNA), and that makes it impossible to draw comparisons with several settlements of the Saltiv culture analyzed by Iuliia Ia. Miagkova.[5]

1 Kroitor 2013.
2 Matolcsi 1984.
3 Miagkova 1998.
4 Koloda and Kroitor 2015; 2016; and 2017.
5 Miagkova 2013.

1 Animal Bones

The animal bone assemblages considered in this book have been discovered on a number of Saltiv sites: Verkhnyi Saltiv (assemblages identified by V. I. Bibikova from the 1959–1961 excavations; those identified by Roman Kroitor from the 1996–1998 and 2004 excavations),[6] Nizhnegnilovskoe,[7] Dmitrievka (identified of Veniamin I. Tsalkin),[8] Karnaukhivka (identified of K. B. Iur'ev),[9] Koropovi Khutory,[10] Lomakino,[11] Mayaki,[12] Mokhnach,[13] Tsimliansk (assemblages identified by A. V. Tattar from the excavations in the 1950s),[14] and P'iatnyts'ke I (especially those assemblages that resulted from the 1987–1988 excavations[15]). Because of serious statistical errors, we have not included in our analysis the animal bone assemblages from Zaplavska (identified by N. M. Ermolova),[16] Netaylivka (identified of V. I. Bibikova),[17] Roganyna, Fashchivka, and Mokhnach P[18] (table 5.1).[19]

A common standard of gauging the importance of animal breeding is the ratio between domestic and wild animals. For example, there is an extremely small share of wild animals in Verkhnyi Saltiv: just specimen animals (2.4 percent), as opposed to 121 domestic animals (97.6 percent). The percentage ratio is 3.6 to 96.4 in Mokhnach; 3.9 to 96.1 in P'iatnyts'ke I; 5.3 to 94.7 in Nizhnegnilovskoe; 6.5 to 93.5 in Mayaki; 8 to 9 in Koropovi Khutory; 9.1 to 90.9 in Lomakino; and 9.5 to 90.5 in Karnaukhivka (Diagram 5.1). Only in Tsimliansk hillfort (at least assemblages resulting from the 1987–1988 excavations) have wild animals reached 41 percent of all animal bone assemblages. However, the materials resulting from Svetlana A. Pletneva's excavations show about quarter of all bone assemblages being wild animals. Be that as it may, with

6 Koloda, Kroitor, and Gorbanenko 2013.
7 Miagkova 1998.
8 Pletneva 1967, 147.
9 Liapushkin 1958b, 313.
10 Kroitor 2013.
11 Miagkova 1998.
12 Matolcsi 1984.
13 Kroitor 2013.
14 Pletneva 1967, 147.
15 Miagkova 1998; Kroitor 2013.
16 Gadlo 1978, 124.
17 Koloda and Gorbanenko 2010, 109.
18 Kroitor 2013.
19 Numbers in Plate 5.1 correspond to those in the diagrams.

 | DOI:10.1163/9789004429574_007

TABLE 5.1 Zooarchaeological materials from Saltiv sites

№	Site	Breeds. amount of individual animals							
		Cattle		Small livestock		Pig		Horse	
		N	%	N	%	N	%	N	%
1	Verkhnyi Saltiv	35	28.9	27	22.3	41	33.9	18	14.9
2	Nizhnegnilovskoe	13	41.9	10	32.3	—	—	8	25.8
3	Dmitrievka	?	29	?	27	?	20	?	24
4	Karnaukhivka	?	34.5	?	28.5	?	14.5	?	22.5
5	Koropovi Khutory*	6	27.3	6	27.3	6	27.3	4	18.1
6	Lomakino	21	47.8	13	29.5	3	6.8	7	15.9
7	Mayaki	22	17.6	61	48.8	27	21.6	15	12
8	Mokhnach	13	48.2	5	18.5	4	14.8	5	18.5
9	Mokhnach P	4	—	1	—	2	—	4	—
10	Netaylivka	8	—	2	—	3	—	3	—
11	Tsimliansk, 1950s	?	39	?	24	?	13	?	24
12	Tsimliansk, 1987–1988	11	24.5	29	64.4	—	—	5	11.1
13	P'iatnyts'ke I	25	25.2	26	26.3	20	20.2	28	28.3
14	Roganyna	1	—	1	—			2	—
15	Zaplavska	4	—	3	—	—	—	2	—
16	Fashchivka	9	—	1	—	—	—	3	—

We calculated all percentages according to data on individual animals of cattle. small livestock. pigs. and horses. For Verkhnyi Saltiv data are combined (see diagrams 5.2–5.5). Because of the inability to determine a reason for discrepancies in data from different excavations in Tsimliansk. they were given and analyzed separately. * Furthermore. donkey.

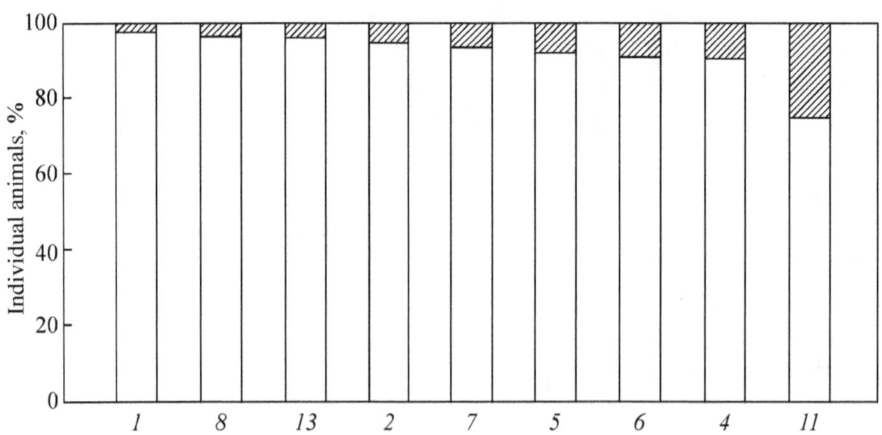

DIAGRAM 5.1
The ratio of wild (above) and domestic (below) animals by quantity of animals (for numbers, see Table 5.1)

the exception of Tsimliansk, hunting seems to have played a minor role in the procurement of meat. Archaeologists and historians have thus concluded that meat was procured mostly, if not only from domestic animals, while hunting was only used for training, for protecting crops against such animals as boars, or for procuring pelts and hides.

Cattle (*Bos taurus* L.) represents about one third of the total number of animal bones in assemblages from Dmitrievka, Verkhnyi Saltiv, Karnaukhivka, Tsimliansk (excavations of Svetlana A. Pletneva), Nizhnegnilovskoe, Koropovi Khutory. In P'iatnyts'ke I, the percentage of cattle varies from 25.2 to 41.9. However, there are cases in which cattle represents less than a quarter of all bones, as

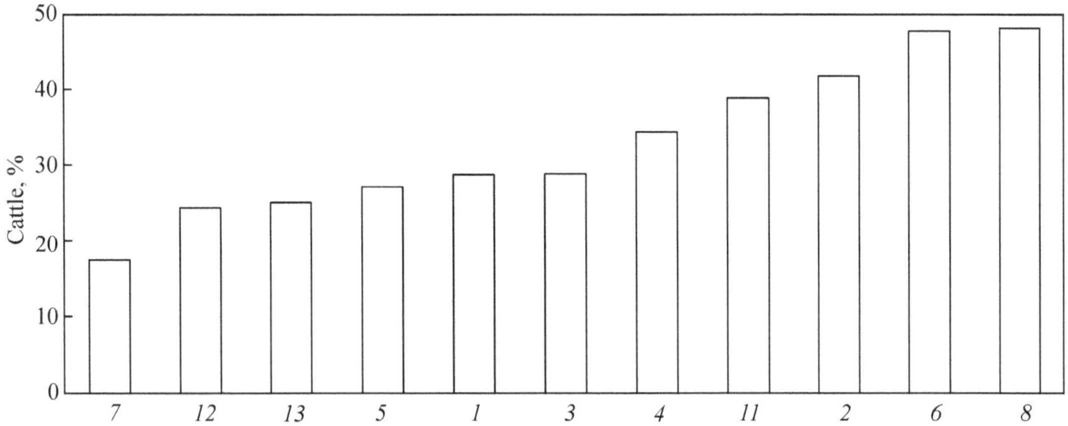

DIAGRAM 5.2
Cattle by quantity (see Table 5.1)

in Mayaki (17.6 percent) or Tsimliansk (the 1987–1988 excavations, from which the resulting cattle bones represent 24.5 percent of all animal bones) (Diagram 5.2). There is therefore a general pattern, which corresponds well to that identified on neighboring Slavic sites (see Chapter 6). For example, according to Anatolii Vinnikovl, between a fifth and a half, but most often between a quarter and third of all animal bones assemblages from sites attributed to the Borshchevo culture are of cattle.[20] Exceptions are Lomakino and the Mokhnach hillfort, where about a half of all bones are of cattle, which is a pattern typical for Rus' towns.[21] Age has been determined for cattle bones from six sites (Table 5.2). The common opinion is that when there is smaller percentage of young animals, animal husbandry well established. That seems to be the conclusion for the Mokhnach hillfort (where 1 in 13 animals was young), Tsimliansk (2 in 11), and P'iatnyts'ke I (5 in 25), but not for Koropovi Khutory (2 in 6) and Nizhnegnilovskoe (6 in 13). In Mayaki, out of 22 animals, 7 were young, 3 semi-adult, and 12 adult. In Lomakino two thirds of the cattle were slaughtered at a young age.

Small livestock consists of sheep (*Ovis aries* L.) and goats (*Capra hircus* L.) (Table 5.2, Diagram 5.3). At Tsimliansk (the 1987–1988 excavations) they are in majority among all domestic animals (64.4 percent), which indicates the presence of the open steppe for grazing flocks. This is further substantiated by the associated bones of wild animals living in the steppe, such as saiga, corsac fox, marmot, and hare.[22] In Mayaki, goats and sheep represent slightly less than a half of all domestic animal bones (48.8 percent). In fact, at a closer look, on most sites, small livestock is between a quarter and a third of all domestic animal bones (Verkhnyi Saltiv, Pletneva's excavations in Tsimliansk, Karnaukhivka, Lomakino, Nizhnegnilovskoie, Koropovi Khutory, and P'iatnyts'ke I). Only in the hillfort of Mokhnach sheep and goats represent less than a fifth of all domestic animal bones. It is worth mentioning at this point that the proportion of small livestock is larger on Saltiv than on Slavic sites. For example, on sites attributed to the Borshchevo culture, the immediate neighbors of the Saltiv sites on the Don, sheep and goats represent only about 15 percent of all domestic animal bones.[23] Similarly, small livestock is about a fifth of all domestic animal bones on sites in Left-Bank Ukraine and along the Donets that have been attributed to the Severians (see Chapter 6).

In Mayaki, sheep prevail within the share of small livestock (38 out of 61 animals). Sheep were slaughtered at a young age more often than goats. Of all the sheep identified within the bone assemblages from the site, 17 were young (45.9 percent), 5 semi-adults (13.5 percent) and 16 adults (40.6 percent), while the age categories for goats had 4 to 3 to 16 individual animals (17.4 to 13 to 69.6 percent), respectively. Judging from animal bones with determined age (Table 5.2), the smallest proportion of small livestock slaughtered at a young age was in Koropovi Khutory (individual animals in proportion of 1 to 5), the largest in Mayaki (21 to 8 to 32).

The horse (*Equus caballus* L.) played a key role in the life and economy of the population of the Saltiv culture (Table 5.1, Diagram 5.4). This can be explained in terms of warfare, farming and tradition. The percentage of horse bones in faunal assemblages varies between 11.1 (the 1987–1988 excavations in Tsimliansk) and 28.3 (P'iatnyts'ke I).[24]

20 Vinnikov 1995, pl. 2.
21 Tsalkin 1956, 143; Potapov 1990, 37–41.
22 Miagkova 1998.
23 Vinnikov 1995, 47 and pl. 2.
24 Within that range, it is worth mentioning the percentage of horse bones in Mayaki (12), Verkhnyi Saltiv (14.9), and Lomakino (15.9).

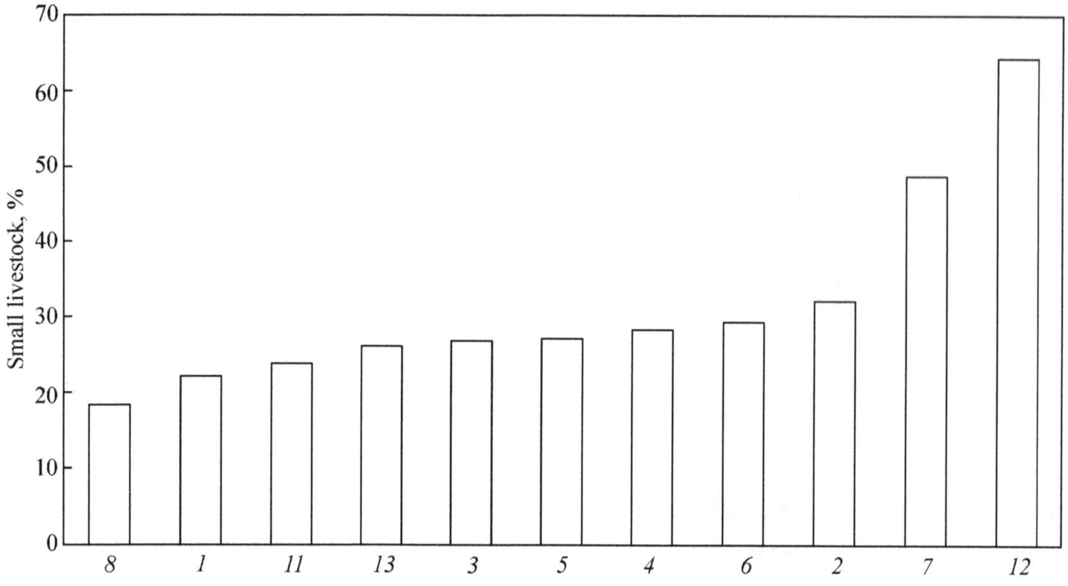

DIAGRAM 5.3
Small livestock by quantity (see Table 5.1)

On the whole, horse bones represent between one fifth and one quarter of all animal bones found in Mokhnach, Koropovi Khutory, Karnaukhivka, Dmitrievka, Nizhnegnilovskoe and during Svetlana Pletneva's excavations in Tsimliansk. On neighboring Slavic sites, the percentage of horse bones is much lower. For example, while on sites attributed to the Borshchevo and Romny cultures, horse bones are few, in general, they represent only 10 percent of all animal bones in the Small Borshchevo, Large Borshchevo, and Titchikha hillforts.[25]

The horses, the bones of which were found in Mayaki were either short (bone assemblage in pit 18; height at the withers is 134.2 cm) or medium-sized (height at the withers varying between 134 and 141.8 cm).[26] This may have possibly inspired Svetlana Pletneva's wrong idea that there were two breeds of horses in Mayaki hillfort: those with long legs, steep croups and small heads, and those with short legs, stocky body, and big heads.[27] Needless to say, there is no basis in the zooarchaeological evidence for any of those characterizations. Most horses died at an advanced age, with only a few foals being slaughtered (Table 5.2). Only in assemblages from Tsimliansk, two out of five horses were foals, and in that respect is quite different from all others.

Pigs (*Sus domestica* Gray) raise a number of very interesting questions (see Table 5.1, Diagram 5.5). First, there were no bones of domestic pigs on two of the eight sites considered for analysis.[28] At Lomakino, the percentage of pigs in bone assemblages in small — 6.8. Elsewhere, the percentage is between 13 and 33.9. To be sure, this is also true for Slavic sites in the Don region, where the percentage varies between 10 and 51.7.[29] By contrast, 30 percent of all animal bones found on sites of the Romny culture are of pig (see Chapter 6). Slightly more than half (55.6 percent) of all pigs from Mayaki died at an advanced age, as opposed to only 37 percent piglets and 7.4 percent semi-adults (see Table 5.2). At Koropovi Khutory, there in six animals were slaughtered before reaching 1 year of age. However, in most other archaeological cultures, the analysis of bones of pig show that most animals were slaughtered at a young age, since the population could be easily reproduced, and its maintenance in summer, unlike in winter, did not require much (if any) labor. That is why we regard the situation in P'iatnyts'ke I to be rather typical (16 to 4 animals died at a young age).

Unlike Slavic sites on the territory of modern Ukraine (which were attributed to the Romny culture), Saltiv sites produced evidence of camels and donkeys. Two camels died in Karnaukhivka.[30] A few bones of camel are also known from sites of the Borshchevo culture.[31] The bones found in Karnaukhivka are to be associated most certainly with the caravan trade, and the same may be true for

25 Vinnikov 1995, pl. 2. Slightly higher percentages are mentioned for the Arkhangelskoe (16) and Zhivotinnoe hillforts (20.7) (Zhuravlev 1998).
26 Matolcsi 1984, 245.
27 Pletneva 1984, 93 with figs. 3–6.
28 However, in bone assemblages resulting from Svetlana Pletneva's excavations in Tsimliansk, pigs represent 13 percent (see Plate 5.1).
29 Vinnikov 1995, 47 fig. 2.
30 Liapushkin 1958b, 313.
31 Vinnikov 1995, 47.

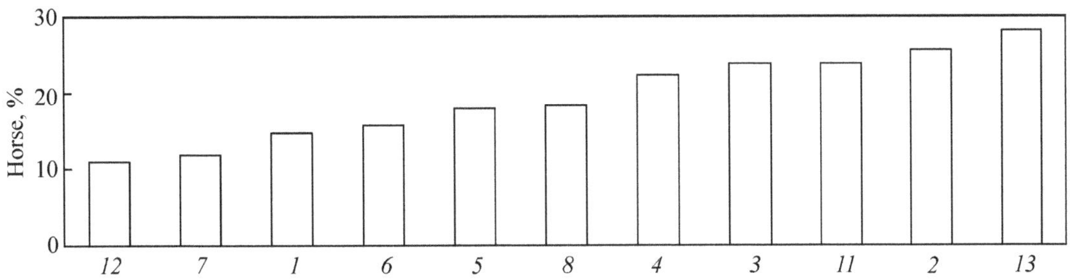

DIAGRAM 5.4
Horses by quantity
(see Table 5.1)

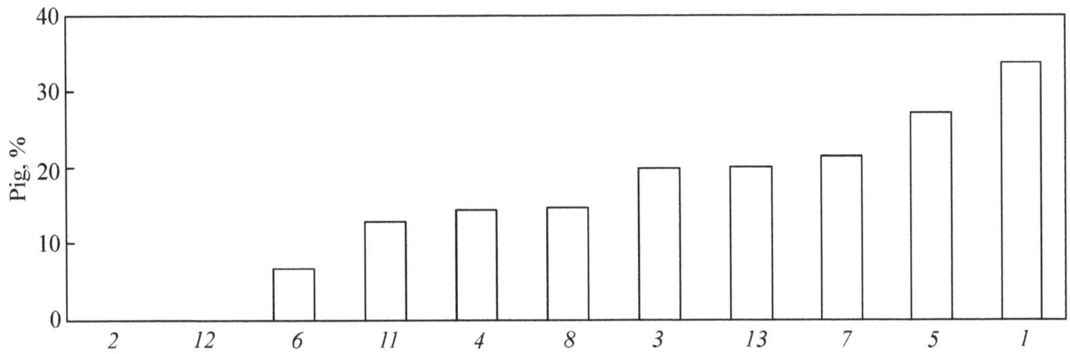

DIAGRAM 5.5
Pigs by quantity
(see Table 5.1)

TABLE 5.2 Correlation of age gradations of domestic animals

Site	Breeds. age ratio											
	Cattle			Small livestock			Pig			Horse		
	Y	S-a	A	Y	S-a	A	Y	S-a	A	Y	S-a	A
Mayaki	31.8 / 7	13.6 / 3	54.5 / 12	34.4 / 21	13.1 / 8	52.5 / 32	37 / 10	7.4 / 2	55.6 / 15	20 / 3	6.7 / 1	73.3 / 11
Tsimliansk hillfort*	18 / 2		82 / 9	39 / 11		61 / 18	—		—	40 / 2		60 / 3
Lomakino	66 / 14		33 / 7	23 / 3		77 / 10	3			20 / 1		80 / 6
Nizhnegnilovskoe	46 / 6		54 / 7	30 / 3		70 / 7	—		—	17 / 1		83 / 7
Koropovi Khutory	2		4	1		5	3		3	1		3
Mokhnach	1		12	1		4	1		3	—		5
P'iatnyts'ke I	5		20	9		17	16		4	6		22

Y — young; S-a — semi-adults; A — adults. In graphs percentage/absolute numbers are given (for the data given by Iu. Ia. Miagkova. we listed them in reverse order. because in most cases compiling percentages is incorrect due to a small amount of data). * Defined by Iu. Ia. Miagkova.

the bones of donkeys found in the Mayaki hillfort and in Koropovi Khutory.[32]

It is also important to take into consideration age correlations across animal species. For example, a great number of the cattle in Lomakino were slaughtered at a young age, which is uncommon. One should keep in mind in this respect that the highest percentage of cattle bones is from that site. It is therefore probable that a cattle breeding was done on this site in order to obtain meat. Cattle were not bred either for dairy or as draft animals. Most of them were slaughtered at the end of the grazing season. Only the reproductive share of the herd was left for the following year. By contrast, in Lomakino small livestock was bred primarily for wool and dairy. In Nizhnegnilovskoe and Tsimliansk, sheep were raised especially for wool, and only secondarily for meat and dairy products. In Mayaki, they were bred especially for quick meat, but in a small number. The latter is indirectly confirmed by the relatively large percentage of pigs in the total number of animals with pigs (while pigs are absent or almost absent on other

32 Matolchi 1984, 238 and 245–246; Kroitor 2013.

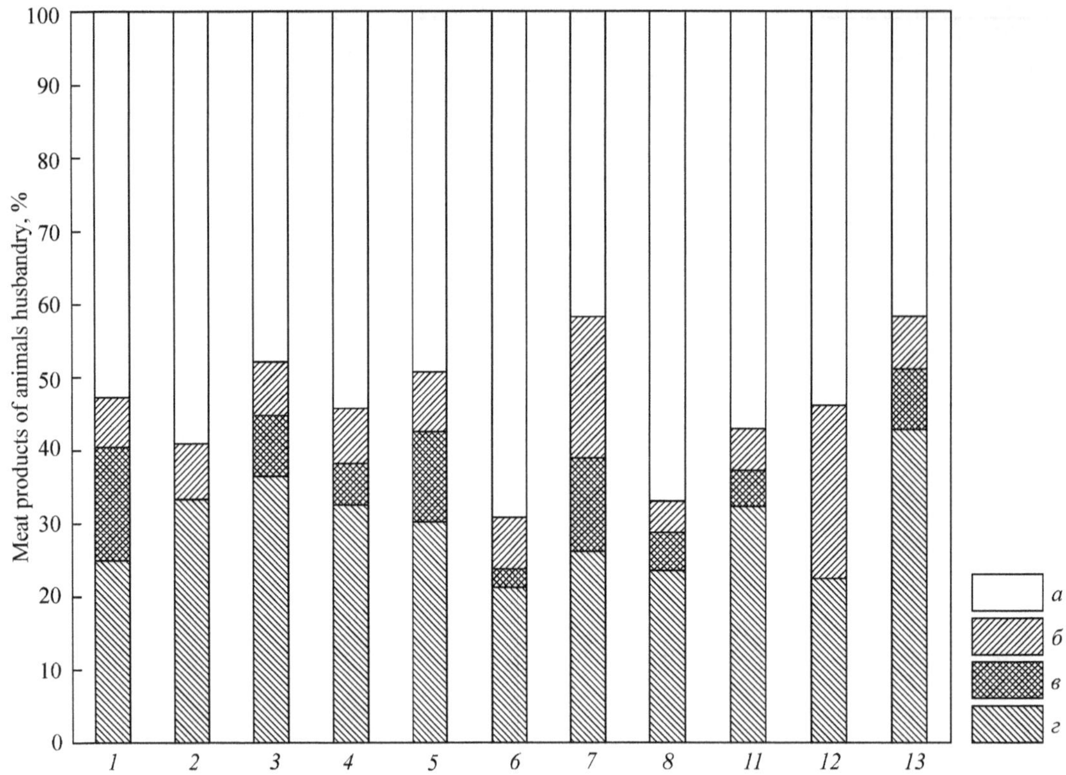

DIAGRAM 5.6 Ratio of domestic animals by quantity of meat obtained (see numbers in Table 5.1): *a* — cattle; *b* — small livestock; *c* — pig; *d* — horse

sites). By contrast, horses do not seem to have been raised (primarily) for meat. Given the rarity of pigs in Lomakino, Nizhenegnilovskoe and Tsimliansk (the 1987–1988 excavations), it may look like pork was not part of the diet on those sites. However, in Lomakino, at least, bone assemblages included remains of wild boars, which increases the amount of pork being consumed to a quantity that is double that established on pig bones alone. In Tsimliansk, the number of specimens of wild board identified on the basis of bones is as large as that of cattle. In Nizhnegnilovskoe, where no pig bones have been found, only one specimen of wild boar was identified.[33] Analyzed in this manner, the data show that the inhabitants of two out of three mentioned sites ate pork; on at least one site they may have been started to breed pigs. Elsewhere (Mayaki, Dmitrievka, Verkhnyi Saltiv, and Koropovi Khutory) the number of pig bones was quite significant. In Netaylivka, three out of 16 domestic meat animals are pigs (Table 5.1). One can safely conclude, therefore, that pigs were bred on Saltiv sites and their inhabitants ate pork. In Mayaki, slightly less than half of all pigs on the site were slaughtered at a young or semi-adult age (see Table 5.2). In P'iatnyts'ke I the ratio between young and adult pigs is 16 to 4. The situation on Slavic sites will be discussed in (see Chapter 6).

33 Miagkova 1998, plate 42.

1.1 *Calculations of Meat Production*

On the basis of the criteria spelled out in Chapter 2, one can attempt to calculate the amount of meat produced on Saltiv sites (Diagram 5.6). A bone assemblage discovered during the 1987–1988 excavations in Tsimliansk will not be considered in this analysis, because it is obviously that hunting played a significant role in meat procurement for the population of the site.

Judging by the number of animals, as well as the largest weight, cattle was the most important source of meat for the population of the Saltiv sites. Cattle dominates unquestionably in that respect on six out of the eight sites considered in this book, on one with 69.2 percent, but on others with percentages around 50. In two cases (the Mayaki hillfort and P'iatnyts'ke I) two fifths of all the meat came from cattle. Horse meat takes the second place, with between a quarter and a third of all meat. The highest percentage for horse meat was recorded for P'iatnyts'ke I, where horse meat was used in the same quantity as beef.

The third and fourth places are occupied by pig and small livestock, but pork has larger percentages than both mutton and goat meat. Nevertheless, no pig bones have been found in assemblages from Nizhnegnilovskoe and Tsimliansk that were studied by Iuliia Miagkova. Small livestock represents between 4.3 and 8.2 percent of all the meat.[34] Higher percentage for mutton, however, have been recorded in Mayaki and in assemblages discovered during the 1987–1988 excavations in Tsimliansk (a fifth of all the meat). Those sites were most likely specialized in sheep breeding.

2 Tools for Animal Husbandry

Animal husbandry is betrayed by finds of stockbreeder tools. In addition to harness elements (the most common), this category of finds includes tools for grazing (for cattle and small livestock) and for shearing (small livestock). In addition, butchering tools provide indirect evidence of animal husbandry. Such tools (butcher knives) are rare, however, perhaps because they were associated with a high degree of specialization. Two tools found in feature 37 in Mokhnach P (Plate 5.1) are most likely butcher knives.[35] They are both long, over 35 cm, with a long blade (23 and 24 cm, respectively). In spite of its length, that blade is straight, as with usual knives. One of the tools has a suspension ring at the end of the handle (Plate 5.1, *2*). Those tools are twice as large as usual, domestic knives, and their handles are bent to the back of the blade, which is slightly curved. The closest analogies are knives for special saddlery leather, the blades of which, however, are twice or three times smaller than those of the butcher knives in question.[36] The most obvious, common feature is the stem bent towards the curved back.[37]

A shank with an attached handle is typical for ordinary knives. Most scholars believe that the handle was made out of a wooden block of appropriate size. In the case of butcher knives, it is obviously, that that was impossible. Most likely, the handle of such a knife was wrapped in coarse fabric (burlap) to absorb moisture during the dissection of carcasses, or in thick leather to enhance the grip on the tool (plate 5.1, *3*). It is unlikely that many more such tools will be found in the future, given that they were not needed in non-specialized contexts. Elsewhere, the dissection of animal carcasses could be done with customary tools. Nonetheless, two analogies are known from Mayaki, others from medieval sites in Bulgaria.[38]

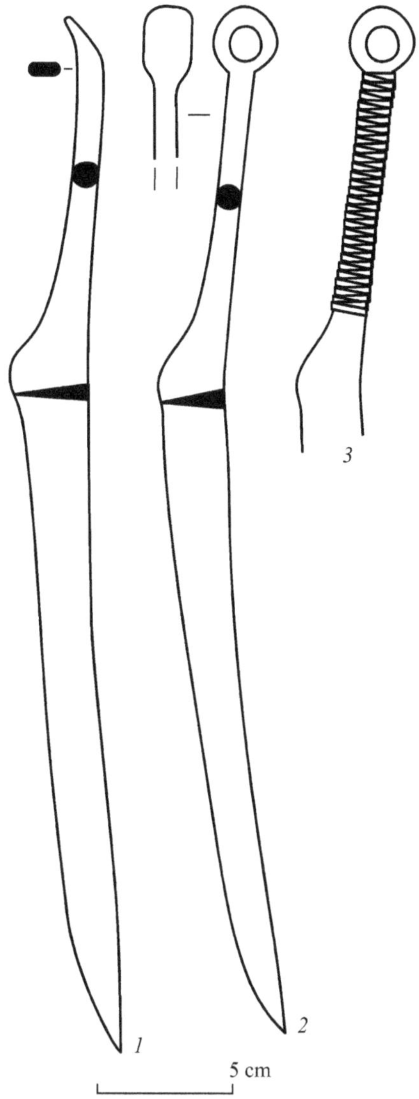

PLATE 5.1
Mokhnach P, feature 37, butcher knives for cutting carcasses: *1, 2* — finds; *3* — reconstruction

Some of the most common finds related to animal husbandry are elements of the horse gear. Most of them come from horse burials or from graves of males buried with horse tack, such as found in Verkhnyi Saltiv and Sukha Gomil'sha.[39] As such, the deposition of elements of horse gear in graves substantiates the idea of horse riding

34 A large number of sheep and goats were slaughtered at very young age, leaving only part of the herd for reproduction. In most cases, the bones of such animals may not have been preserved at all; especially the meat (with the bones) was cooked. Dogs may have also been fed with such bones.

35 Koloda 2015.

36 Knives believed to be for saddlery leather have been found in Mokhnach P, the site of origin for the two butcher knives (Koloda 2017).

37 Koloda 2014, 71, fig. 3, *11, 14*.

38 Mikheiev 1985, fig. 19, *17, 18*; fig. 32, *1–3*.

39 Mikheev 1986a; Aksenov 2001–2002, 270 and 276 figs. 2–6; Koloda 2004, 221–225. As both cemeteries are characterized by various burials customs, the general presence of elements of horse gear underscores the significance of the horse for ethnic groups making up the population of the Saltiv archaeological culture.

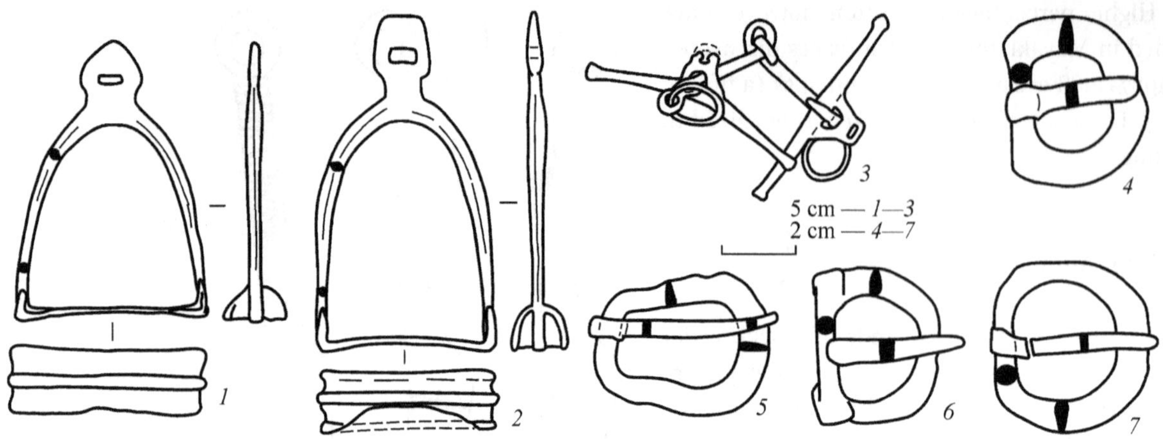

PLATE 5.2 Elements of horse gear from settlement sites: *1, 2* — stirrups; *3* — bridle-bits; *4–7* — horse tack buckles (*1, 2* — Mokhnach P; *3–7* — Mokhnach)

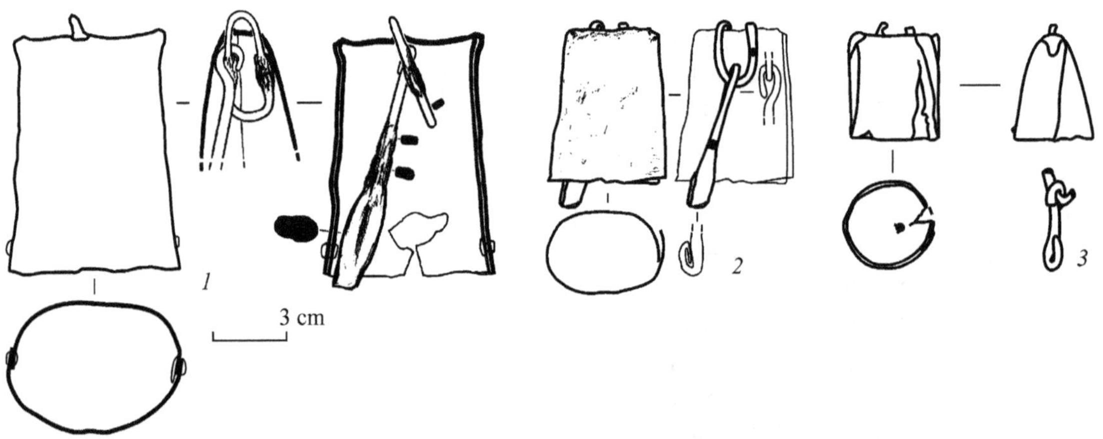

PLATE 5.3 Bells for livestock: *1* — Kochetok; *3* — Mokhnach

having a crucial significance in Saltiv communities. Given that most burials assemblages with elements of horse gear also included weapons, one can draw the conclusion that horse riding was primarily linked to warfare. Several studies have so far been dedicated to the typology of horse-gear elements, as well as weapons found in burial assemblages, to their number and social value.[40] Comparatively less attention has so far been paid to finds of horse-gear elements from settlement sites or other non-funerary assemblages. In the region under study, elements of horse gear are known both from isolated (stray) finds,[41] and from settlement features.[42] Judging by so many finds, the horse equipment in Khazaria was quite standardized, to such a degree that it has now become a crucial marker of cultural affiliation for new sites under excavations. For example, the Mokhnach site produced bridle bits with vertical, bar-like cheek-pieces (Plate 5.2, *3*), stirrups of different types (Plate 5.2, *1, 2*), and horse tack buckles of different shapes (Plate 5.2, *4–6*). By contrast, there are far fewer tools for animal husbandry (which refer to animals other than horses). This fact is important to mention, as both sheep and cattle bells are quite common on Saltiv sites.[43] Another bell, much poorly preserved, is known from Mayaki, where sheep breeding is clearly documented by zooarchaeological data (see above).[44] In the Donets area, similar bells are known from the Mokhnach hillfort, and from one of the cremation cemeteries near Kochetok (Plate 5.3).

Shears constitute the strongest evidence that small livestock was bred not only for meat, but also for wool. The fragmented of one specimen was found in the Verkhnyi Saltiv hillfort,[45] two others in Koropovi Khutory and in

40 E.g., Kryganov 1989; Aksionov 2000 and 2005, pp. 61–65, figs. 1–3.
41 Gorbanenko and Koloda 2010, 36 fig. 10; Koloda and Kvitkovskii 2008, fig. 5.
42 Koloda 2012, 35; Koloda 2015, 113–114.
43 Mikheev 1985, 27.
44 Vinnikov and Pletneva 1998, 186, fig. 74, *D*.
45 Gorbanenko and Koloda 2010, 36 fig. 11.

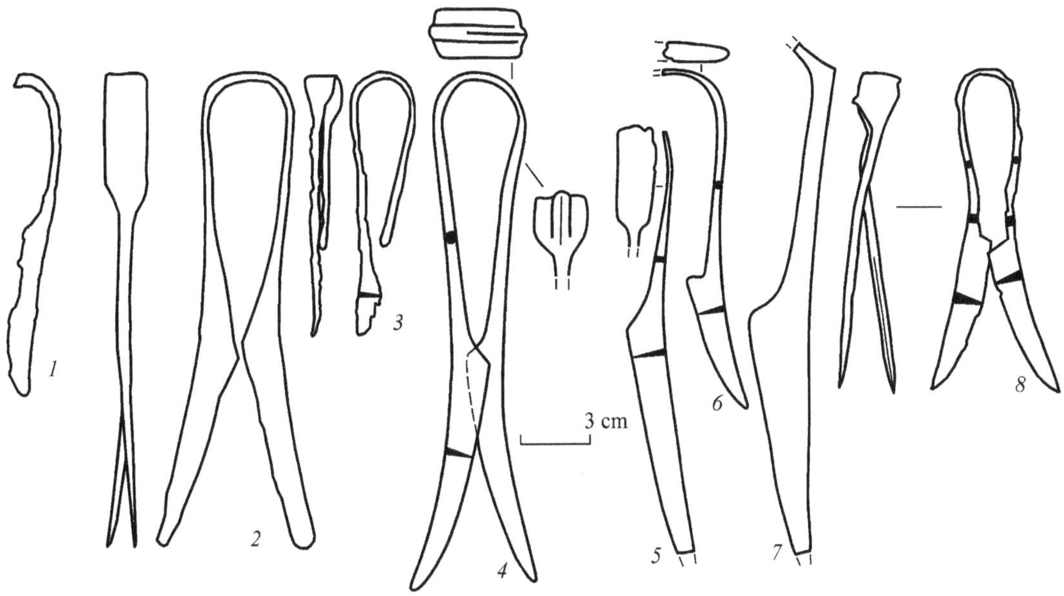

PLATE 5.4 Shears: *1* — Verkhnyi Saltiv; *2* — Koropovi Khutory; *3* — Kochetok; *4–7* — Mokhnach P; *8* — P'iatnyts'ke I

P'iatnyts'ke I. Another fragment was found in Kochetok.[46] Four shears in different forms of preservation have been found in Mokhnach P (Plate 5.4). Such tool were quite common in the southern parts of Eastern Europe ever since the beginning of the first millennium AD. Spring shears are known from many Saltiv sites[47] and from contemporaneous Slavic sites as well.[48]

3 Livestock Maintenance

If applying K. P. Buniatian's criteria, animal husbandry on Saltiv sites is of two kinds — on one hand, stall-and-pasture, on the other hand transhumance.[49] The system of stall and pasturable animal husbandry required pens and other shelters for animals. Stall-and-pasture means that the cattle spend the day grazing in the fields, but return to the settlement for spending the night in stalls and other facilities for keeping livestock. This type of animal husbandry is associated to paddocks, the remnants of which may be found inside settlements or next to them. The stall-and-pasture system of animal husbandry is a correlate of developed soil cultivation.[50]

A necessary condition for the stall-and-pasture system is the availability of grazing fields located within easy reach. In some cases, the fields left fallow to regain soil fertility may be used for that purpose, in other cases, the edges of forest areas, which are not suitable for soil cultivation or have not yet been cleared. Finally, neighboring floodplains that are otherwise not good for soil cultivation for various reasons (excessive moisture, propensity to flooding, relief features) can serve as pastures.

Enclosures (perhaps with awnings) for keeping livestock during the warmer months most likely existed near the pastures, much like nowadays. An opened paddock for livestock required posts (each 2 m long, and 10 to 15 cm in diameter), planted every 2 to 3 meters along the perimeter. Transverse poles were then attached to those posts at the height needed.[51] That much is known from the ethnographic evidence. As far as the archaeological evidence is concerned, the existence of paddocks for cattle has been documented for 10th- to 13th-century Rus' settlements.[52] Of particular interest is feature 195 in Avtunichi, because of its graphic reconstructions. The structure had an irregularly quadrilateral shape with 19 postholes, enclosing a 7 by 5 m area.[53] Besides that, the evidence for paddocks is extremely rare, largely because of the location. Often out of the settlement area, where archaeological investigations are hardly carried out.

The other component of the stall-and-pasture form of animal breeding are the structures meant to provide shelter for the livestock during the cold season.[54] One

46 Svistun 2012, fig. 6, *44*.
47 Mikheev 1985, 27, fig. 11, *17*.
48 Vinnikov 1995, fig. 17, *1*; Kovalenko 1999, 39 fig. 7.
49 Buniatian 1997.
50 Buniatian 1997, 33.
51 Bondarchik 1979, 23.
52 Motsia 2003, 176.
53 Gotun 1993, 69 and 70.
54 Liapushkin 1958b, 269; Vinnikov 1984, 122; Krasilnikova 2005, 11. Equally relevant are the materials in the field report of K. I. Krasil'nikov's 1976–1977 excavations.

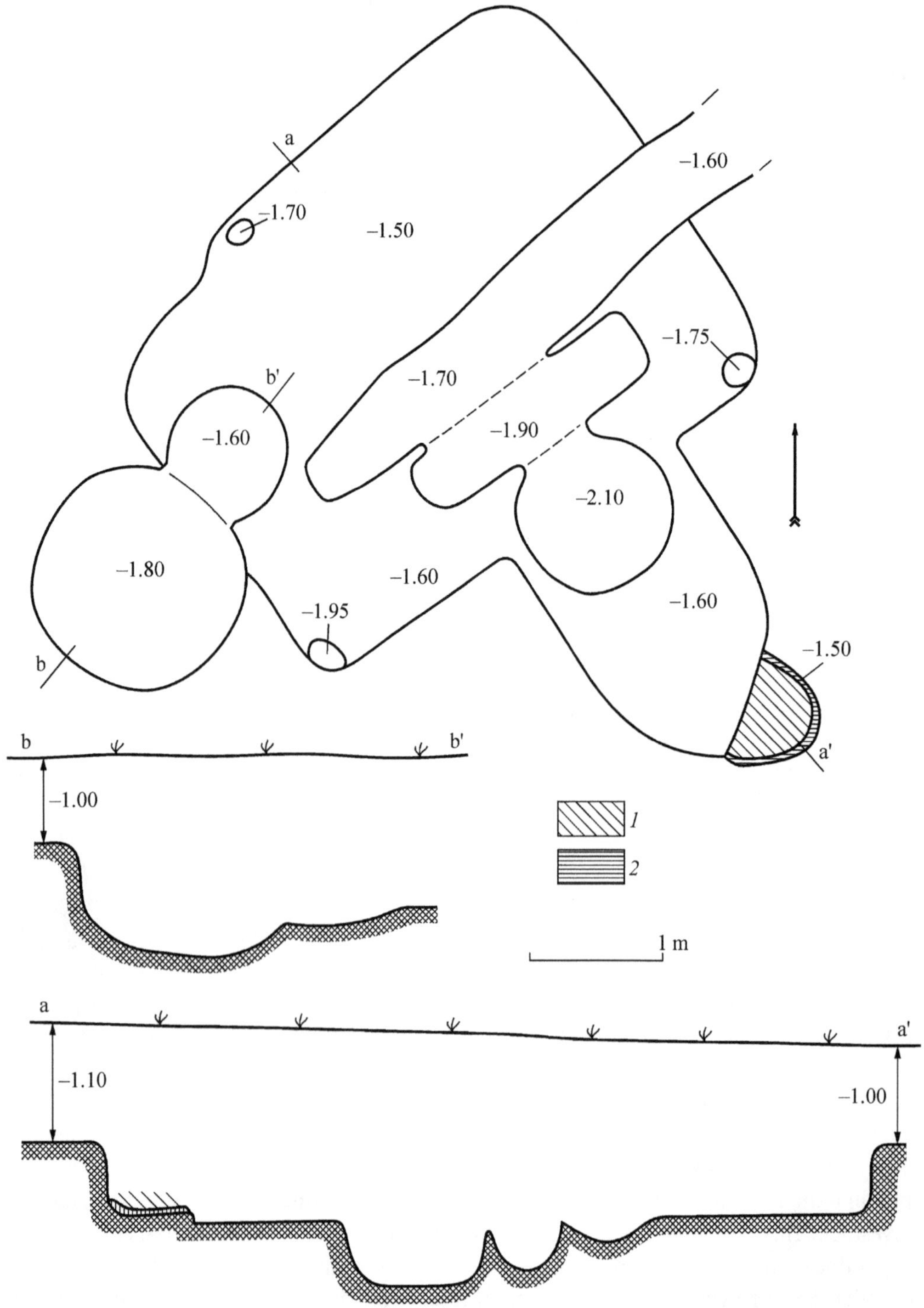

FIGURE 5.1
Verkhnyi Saltiv, animal stall: *a* — ash and charcoal, *b* — burnt clay

such structure, rightly interpreted as a stall, was excavated in 1998 in Verkhnyi Saltiv.[55] The structure had a quasi-rectangular plan (2.70 to 3.20 by 5.20 m), 1.50 to 1.60 deep in relation to the modern ground (Fig. 5.1). Postholes with a diameter of 20 to 25 cm, each between 1.70 to 1.95 m deep, were found in the middle of the northwestern side, as well as in the eastern and southern corners. There was also a double, 8-like pit on the southwestern side of the building. Its northern part had a diameter of 90 cm, and was 1.60 m deep. The southern part of the pit, going beyond the structure, was 1.50 m to 1.60 m long and 1.80 m deep. There was a niche in the southeastern side, 1.50 m wide and 2 m long. An oval fireplace (70 by 65 cm) was found inside the structure, the hearth of it was strongly calcinated. In front of the niche was a round pit (2.10 m in diameter), with a flat bottom, 1.10 m deep. A 1.60 to 1.70 m deep groove was in the central part of the structure but ran beyond the room to the north-east for a total (examined) of

55 Gorbanenko and Koloda 2010, fig. 12.

7 m. Another rectangular deepening (2 by 0.5 m) was discovered between the groove and the fireplace. It was 1.90 m deep. This structure cannot be interpreted as a house (dwelling). The opened hearth inside the deep niche is unique for the housing construction of the Saltiv culture, and has no parallels anywhere else in Eastern Europe. Nor can the fireplace have offered comfortable temperature for human living. The 8-like shallow pit was also not fit for food storage. The same is true for the pit near the southeastern wall. Twin grooves of considerable length and width in the center of the floor must also have been extremely inconvenient for human habitation. This, in other words, was most likely a stall equipped with a simple heating device, located as far as possible from the livestock in the main room, and partitioned by the manger pit. One more pit of similar purpose was found near the south-western wall. Its 8-like shape, based on its profile, may be associated with a partial trampling and further deepening of the floor caused by livestock. The grooves were probably for draining animal waste outside the stall. The wide and shallow drainage groove coming outside the structure was probably the entrance to the stall (especially since it opened towards a floodplain meadow).

A room with drainage grooves of large size and depth has also been found in Karnaukhivka.[56] Two other, similar structures are known from the Mokhnach hillfort. One of them was part of a blacksmith's house (Fig. 5.2, *1*).[57] This was a rectangular structure (2.20 to 2.60 by 4.40 m) with greatly rounded corners. Its bottom inclined towards the southwestern edge was found at a depth of 65 to 90 cm. An oval niche (30 by 50 cm) was found in the northwestern corner, at a depth of 50 cm. The entrance was on the northern side, 90 to 1.20 m large, with a 30 cm-wide step. A 3.70 m-long and 20 to 40 cm-wide groove parallel to the wall was found in the western part of the structure. In the center was a shallow pit of oval plan (75 by 80 cm), 10 cm below the floor of the structure. Besides fragments of Saltiv pottery, its black soil filling contained several pieces of sandstone. The interpretation of this structure depends upon the understanding of its plan, interior arrangement and insignificant depth. The stratigraphic observations, conducted in this part of the site, indicate that the walking level of the Saltiv settlement was 25 to 30 cm lower than the modern one.[58] This means that at the time it was in operation, the floor of the structure was at a depth of 35 to 60 cm, which is significantly less than all sunken-floored house known from this site. Moreover, the absence of any fireplace makes it unlikely that the structure served as a dwelling. A drainage groove of such a large size was neither needed, nor indeed useful either in a house or in a workshop. The central pit, as well as the niche in the corner, may have been a manger, similar to that complex in Verkhnyi Saltiv. A conveniently slight slope of the floor was no problem for all types of ungulates. The presence of the outer entrance hall (corridor) may be interpreted as an addition designed for maintaining the heat inside the structure.

Another stall was located in the northern yard of the hillfort, and was excavated in 2016 (Map 1.7). This was also a rectangular structure (2 to 2.20 by 3.50 m) with rounded corners and a slight enlargement in the southern part. Its flat bottom was found at a depth of 70 cm. Steps were found near the south-western corner at a depth of 50 cm. Almost in the center, and across the entire structure, there was a 90 cm-long groove (0.30 to 0.35 by 3.10 m). The groove was most likely designed for drainage of the animal waste. The filling of the structure contained very few artifacts, but a heap of fired clay coating and pieces of charcoal were detected along the eastern side. This may have been a fireplace meant to warm up the stall in the winter (Fig. 5.2, *2*).

Besides the stall-and-pasture system of grazing, transhumance must also have been practiced, especially by mobile groups of residents in settlements.[59] The analysis of potential resource zones of the Saltiv sites in the Donets region (the Mokhnach hillfort, Koropovi Khutory, and Verkhnyi Saltiv) shows that grazing fields were available at a certain distance from each one of them. For example, the pasture was 2.5 km south from the Verkhnyi Saltiv settlement, and was clearly isolated by the natural boundaries — the Donets to the east, a small ravine with a stream to the south, and another ravine towards the settlement. In other words, the shepherd(s) bringing the animals to this pasture needed to guard it only on the northwestern side. The left bank part of the potential resource zone also offers grassland for transhumance or for keeping the livestock in stall over the winter (see Map 3.2 in chapter 3).

Such pastures were characteristically located at the foot of the hill on which the Mokhnach hillfort is located, with an extension to the north, on the eastern side of the hillfort, towards the bends of the Donets. Given its isolation, this area may have well been used for animal husbandry

56 Liapushkin 1958b, 269.
57 Koloda 2002b; 2005, 162–163.
58 As Afanas'ev 1987, 51 notes, such a stratigraphic relation is confirmed by data from other Saltiv sites in the forest-steppe belt.
59 Buniatian 1997, 34–35.

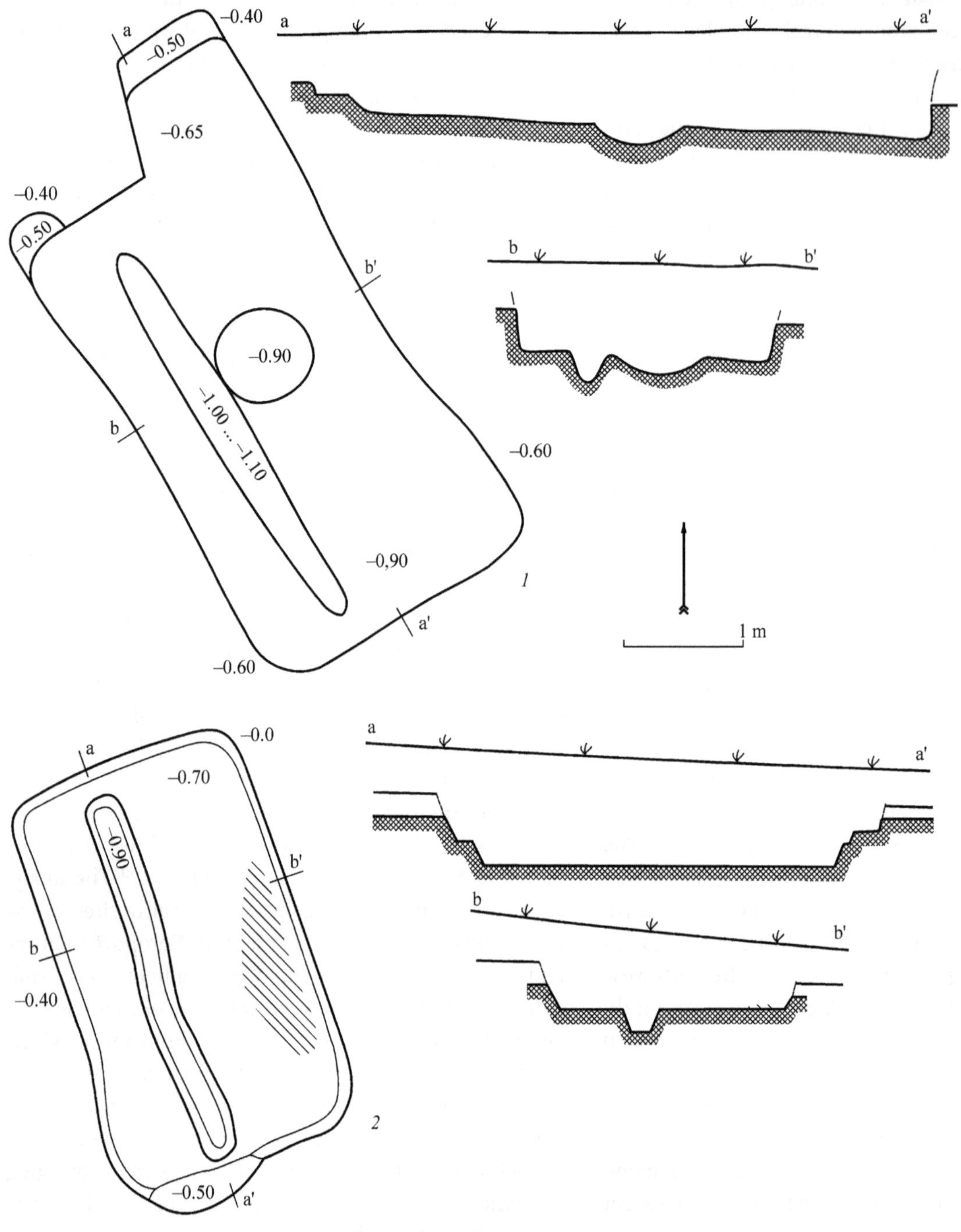

FIGURE 5.2
Mokhnach, animal stalls: *1* — feature 1; *2* — feature 99

(Map 3.4).[60] Such conditions perfectly met the economic needs of the medieval population. One can go so far as to suggest that those conditions were crucial in the selection of the site, particularly in light of the need to monitor the grazing animals. It may be significant to mention at this point that in the early Middle Ages (between the mid-8th and the mid-10th century), the Mokhnach hillfort covered an area of about 12.5 ha (31 acres), with an agglomeration of about 100 co-existing residences. Moreover, at that same time, there were at least 14 settlements in the hillfort's hinterland (see Chapter 1).[61] All data indicate that during the Saltiv period the hillfort in Mokhnach hillfort was a center of a significant agricultural region.

Plots of land ideal for animal husbandry have also been identified in the hinterland of Koropovi Khutory, on both

60 It should be noted that the oxbows that are now in existence, and are fueled by springs arising out of the hill with the hillfort, may have at one time have been part of the riverbed. In addition, it is probable that the site was divided by water flowing into large areas. That much results from the presence of contemporary additional channels in the southern part of the Mokhnach village.

61 Koloda and Koloda 2001, 43–44; Koloda 2010a; 2010b; 2012b.

sides of the river. They were also neatly delineated by natural boundaries, and could thus be used for transhumance or/and for hay-making meant for keeping livestock in the stall over the winter (Map 3.5). A similar plot of land may be distinguished near the P'iatnyts'ke I settlement. On the outskirts of the potential resource zone of that site, to the north there is aa partially isolated area, which is ideal for the transhumant grazing system because of natural boundaries delineating the area (Map 3.3). This, however, does not preclude grazing on any other areas around the settlement, if they were not used as the fields.[62]

Grazing may also be associated with occasional finds of dog bones. No less than 5 individual animals were identified in from Verkhnyi Saltiv, another in Netaylivka.[63] Small numbers of dog bones are known from other sites as well.[64]

A number of conclusions may be drawn from this survey of animal husbandry.

A direct comparison between wild and domestic animals identified in bone assemblages shows beyond any doubt that hunting ha an insignificant role in the meat procurement for the sedentary population of the Saltiv culture. Developed animal husbandry entirely satisfied the needs for meat. The only exception is the materials resulting from the 1987–1988 excavations in Tsimliansk. Cattle were the predominant species among domestic animals, as well as the main source of meat. It is however clear that cattle were the main draft animals for arable tillage. In most cases, small livestock play a secondary role (if any) in meat production. Sheep were definitely raised more for wool (as indirectly confirmed by finds of spring shares), than for fresh meat. There were however sites that specialized in sheep breeding. A large number of horses (much larger than on contemporaneous Slavic sites) indicates the important role of horse riding in the Saltiv culture. It is of course quite possible that horses were also used as draft animals for soil tillage, particularly for plowing light or old-arable soils. Moreover, on Saltiv sites, horse meat ranks second only to beef. Pigs, however, were not raised everywhere. On sites located in the forest-steppe zone (Verkhnyi Saltiv, Dmitrievka, Koropovi Khutory, Mayaki, Mokhnach, and P'iatnyts'ke I) the percentage of pigs within the total number of animals is significantly larger than on sites in the steppe (Karnaukhivka, old excavations in Tsimliansk). In fact, on some sites in the steppe zone, such as Nizhnegnilovskoe, Zaplavska, and Fashchivka, there are no pig bones at all. Pork, therefore, played a different role in the diet on various sites. This variation may have something to do with natural and climatic conditions. The animal husbandry model in the steppe zone was nomadic or semi-nomadic. The sedentary population of the forest-steppe zone practiced both an advanced form of farming and animal husbandry. Pigs were therefore much more adapted to a sedentary form of life. It is possible, however, that differences in the role of pork in the local diet have something to do with specialization in animal breeding, ethnicity or even the religious affiliation of certain groups of the Saltiv population.

Notable differences in animal husbandry between the forest-steppe zone of Khazaria and the neighboring Slavic region are marked by small numbers of camels and donkeys. Both are most likely the result of Eastern influences, namely of the participation in the caravan trade along the Great Silk Road.[65] One should not discount the exotic aspect of the presence of such animals in Khazaria, which could have been interpreted symbolically.

Grazing is indicated by finds of dog bones and of bells for livestock. Livestock grazing was most likely of the stall-and-pasture type, in which the livestock returned to the paddocks for part of the day. This grazing system is indirectly indicated by places suitable for grazing that are located just near the sites. Transhumant grazing was also possible. That system presupposed grazing in remote areas (most often in the summer) and returning livestock to settlements for some part of the year (usually autumn and winter). An analysis of the territories around the settlements indirectly confirms this opinion. The presence of stalls, some of them heated, is confirmed archaeologically.

Plant cultivation and animal breeding seem to have complemented each other quite well. Crops may have cultivated primarily (if not only) to feed the animals (straw, oats barley, rye; as well as barley and oats grain). The level of both plant cultivation and animal husbandry was quite advanced in the Saltiv culture. The indirect evidence of that is the rather insignificant of hunting in meat procurement, the artifacts associated with livestock breeding, the artifacts related to animal husbandry, and the organization of the territory around the settlements.

62 To this day, the territory of the site, as well as the floodplain of the Velyka Babka river floodplain between the modern-day villages of P'iatnyts'ke and Kytsivka (to the south) are used by local residents for transhumant grazing in the summer.

63 Koloda, Kroitor, and Gorbanenko 2013; Koloda and Gorbanenko 2010, 109.

64 Liapushkin 1958b, 313; Matolcsi 1984, 238 and 258; Miagkova 1998.

65 Pletneva 1996, 142–158.

CHAPTER 6

The Khazars and their Neighbors: a Comparative Analysis, in Lieu of Conclusions

To assess the level of development in the agriculture practiced by Saltiv communities in the Donets area, one needs to consider it within a broader geographical context. For that purpose, similar data from "neighboring" archaeological cultures are used, primarily those from sites to the north attributed to the Borshchevo culture on the Don and to the west and northwest attributed to the Volyntseve and Romny cultures. Materials from sites of the Raiky culture are rarely used for comparison, because there were no direct contacts between Saltiv and Raiky communities (see Map 1.2). Given the similar natural conditions of the East European forest-steppe in which they developed, the Saltiv and Romny cultures have much in common in terms of agriculture. However, there were also substantial cultural (and ethnic) differences. The constant mutual influences, mostly of a pragmatic and positive nature, resulted in much sharing of technology and innovations. This is otherwise expected in borderlands and contact zones like that under investigation in this book.

For a meaningful comparison, we have dealt separately with arable farming and animal husbandry. We have also adopted the following order: zones around settlements (for both branches of agriculture); tools for farming, palaeobotanical spectra, crop storing and system (for soil cultivation);[1] zooarchaeological spectra, tools for animal husbandry, animal housing (for animal husbandry).

1 Zones around Settlements

The zones around the settlements of the Saltiv and the synchronous Slavic cultures in the contact region are quite similar. According to our own observations on a large body of material pertaining to the archaeological cultures attributed to the Slavs, at the end first millennium AD settlement siting was primarily based on the needs of agriculture.[2] In other words, the Slavs were quick to adapt to the natural environment. Obvious examples of that adaptation are the sites near Volyntseve (Sumy region, Ukraine),[3] as well as the hillforts in Sverdlovka 1 (Chernihiv region, Ukraine),[4] and Zhivotinnoe (Voronezh region, Russia).[5] Next to each one of those sites, several areas existed, some good for soil cultivation, others for animal husbandry. The detailed investigation of the zones around those settlements shows a sophisticated and quite flexible use of the environment in order to maximize the output of agricultural production. In that respect, there seems to be no difference between the Saltiv and the Slavic sites. However, on the latter there are supposedly no signs of slash-and-burn, as a first phase in gaining more fields for cultivation. In other words, Slavic sites had presumably more land available for cultivation that did not require clearing.[6] This is clearly true for the later, "Slavic" phases of occupation on such Saltiv sites as Mokhnach and Koropovi Khutory.

2 Arable Farming

Tools for soil cultivation have been found on sites attributed to the Borshchevo and Volyntseve-Romny cultures (Plate 6.1–6.3), as well as to the Raiky culture (Plate 6.4).

To be sure, metal parts of tools for soil tillage are present in all cultures dated to the last quarter of the first millennium, but not in the same numbers. There is a very large quantity of such artifacts in assemblages of the Saltiv culture.[7] Fewer than ten coulters have so far been found on sites of the Volyntseve-Romny culture, and about 30 plowshares from eight sites (Plate 6.2).[8] In the Borshchevo culture, there is only one fragment of a

1 We somewhat deliberately retreated from reviewing the materials of soil cultivation according to the agricultural process, to be able to analyze similar materials together (artifacts — PBS — objects attributed to crop storing — soil cultivation system).

2 Gorbanenko 2007, 93; Gorbanenko 2009, 98 and 99; Gorbanenko and Pashkevich 2010, 276.

3 Gorbanenko 2003–2004. Some of those sites have been attributed to the Volyntseve culture (named after the eponymous site), others are dated to the Rus' period.

4 Chernenko et al. 2017. The hillfort in Sverdlovka 1 has been attributed to the Romny culture.

5 Vinnikov and Gorbanenko 2013. The Zhivotinnoe hillfort has been attributed to the Borshchevo culture.

6 Gorbanenko and Koloda 2013, 69–74.

7 A significant number of finds from the Donetsk region of Ukraine still remain unpublished.

8 Veretiushkina 2008 and 2011; Gorbanenko and Pashkevich 2010; Gotun and Gorbanenko 2016.

 DOI:10.1163/9789004429574_008

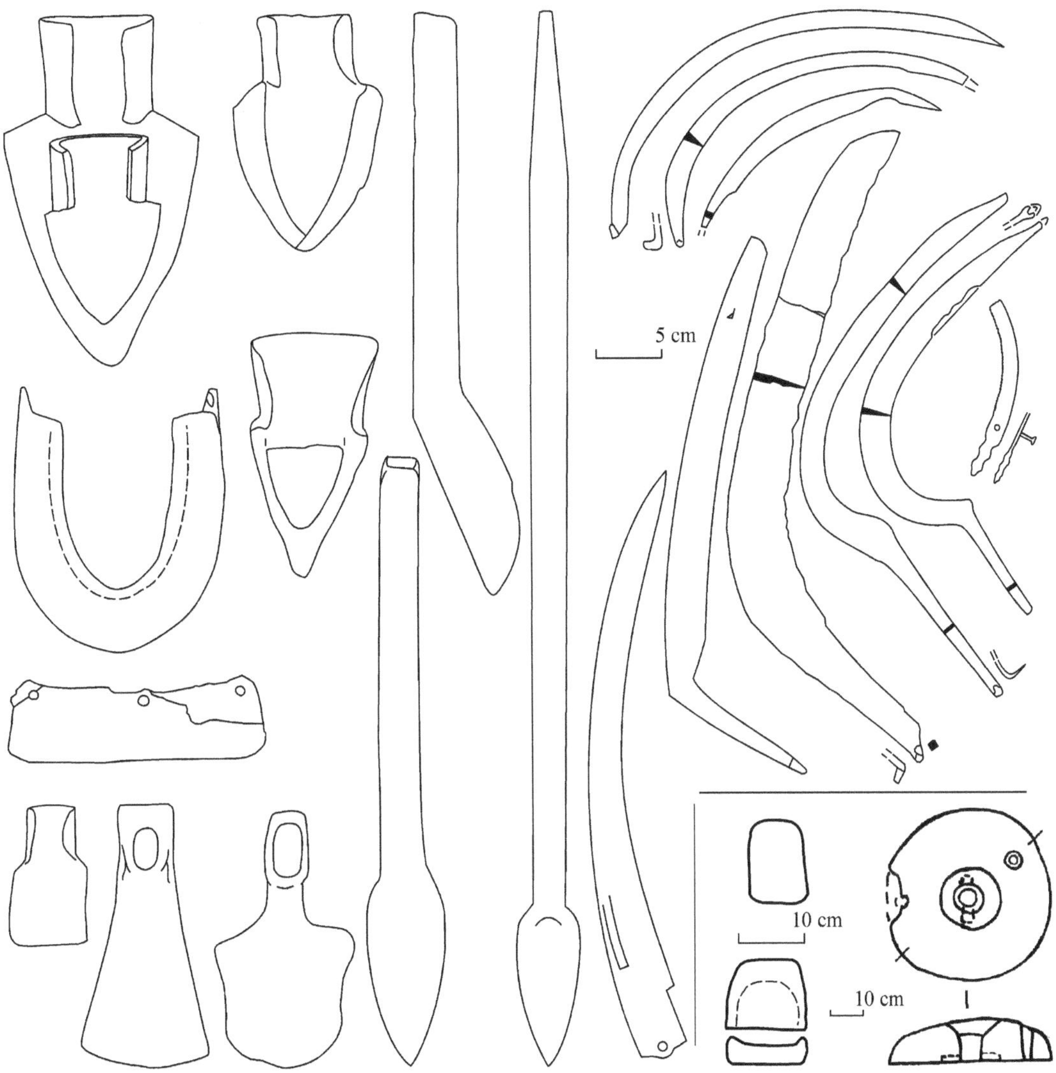

PLATE 6.1
Details of agricultural tools from Saltiv sites

supposedly a narrow-bladed plowshare.[9] It is quite probable that socketed coulters, which seem to be typical for assemblages attributed to the Slavs, were also in use in the Don area (Plate 6.3).[10]

Raiky sites have produced plowshares of different types: type I A 1 is documented in the Pastyrs'ke hill for (Cherkassy region, Ukraine), but there are also specimens of types I B 1 and I B 2. Coulters are also known from Raiky sites. To this day, 13 plowshares and 5 coulters have been found on sites attributed to the Raiky culture. An additional plowshare comes from Boiarka (Kyiv region, Ukraine). All those artifacts may be dated to the last quarter of the first millennium.[11]

On balance, it appears that Saltiv sites have produced a greater quantity and variety of arable tool parts (see Plate 6.4). In that respect, the Saltiv culture is clearly superior to those in the immediate neighborhood, and the situation is unlikely to change in the nearest future due

9 It is important to refrain from hastily drawing conclusions about the level of economic development in the Borschevo archaeological culture, particularly of its agriculture. In the mid-1920s, Petr M. Tret'iakov insisted that slash-and-burn was typical for the agriculture practice by Slavs in Left-Bank Ukraine during the last quarter of the first millennium. His ideas were based on the extremely poor material resulting from the excavation of sites of Romny culture. He drew the conclusion that a shift to arable farming among those Slavs could not have taken place before the end of the first millennium (Tret'iakov 1946, 43; Tret'iakov 1947, 131–133; Tret'iakov 1951, 49–56). However, later excavations on such Volyntseve-Romny sites as the Novotroits'ke hillfort (Liapushkin 1958c), Bytytsa (Liapushkin 1959–1960), and Volyntseve (Dovzhenok 1952) produced an abundant material pertaining to the development of agriculture. As a consequence, Tret'iakov changed his views, and began advocating the idea that arable farming was already in place in the middle of the first millennium AD (Tretiakov 1966, 301; Tret'iakov 1969, 22 and 25). Even that may be subject to change in the future, especially in relation to the Don region.

10 Kovalevskii and Gorbanenko 2014.

11 Gorbanenko and Pashkevich 2010; Mykhaylyna 2014; Pyvovarov and Il'kiv 2014; Gotun and Gorbanenko 2016.

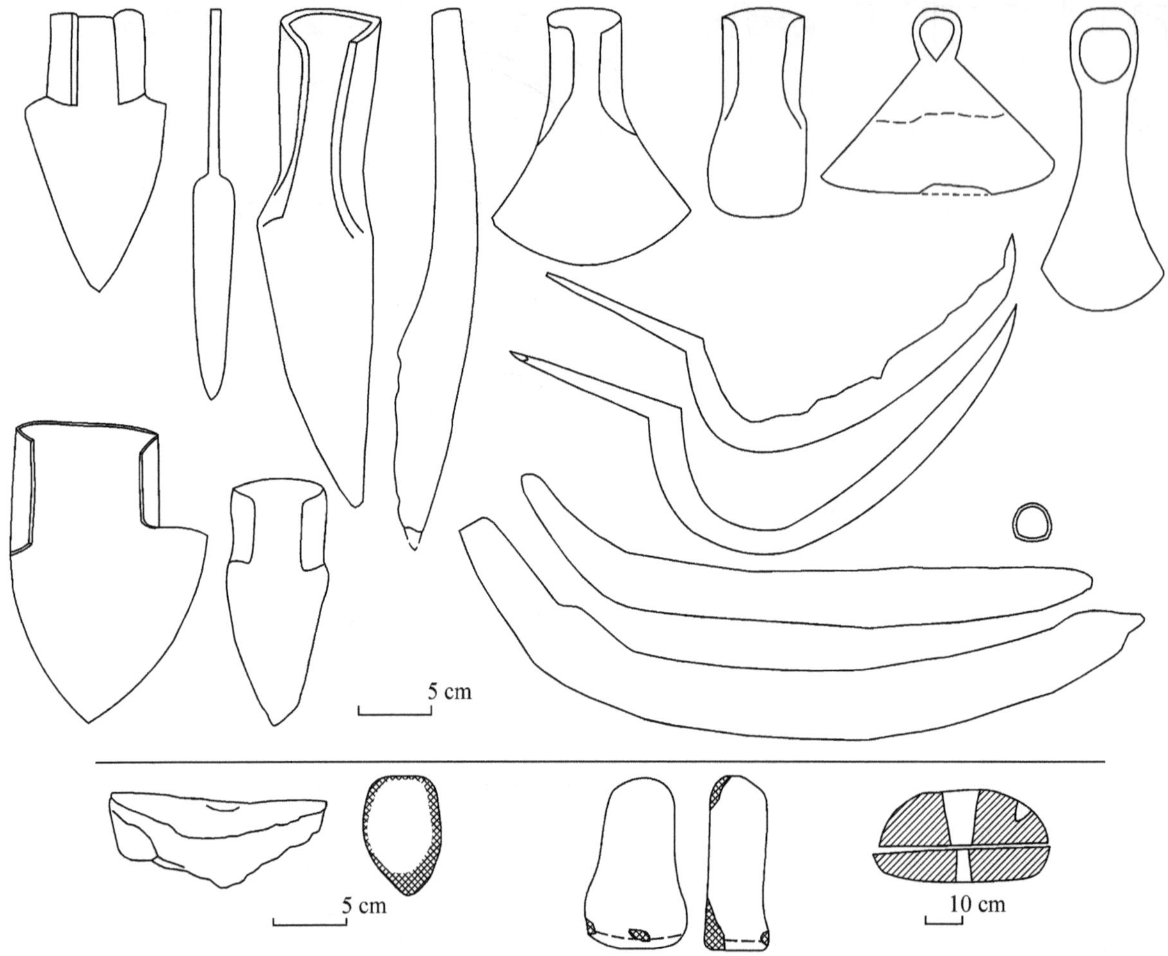

PLATE 6.2 Details of agricultural tools from Volyntseve-Romny sites

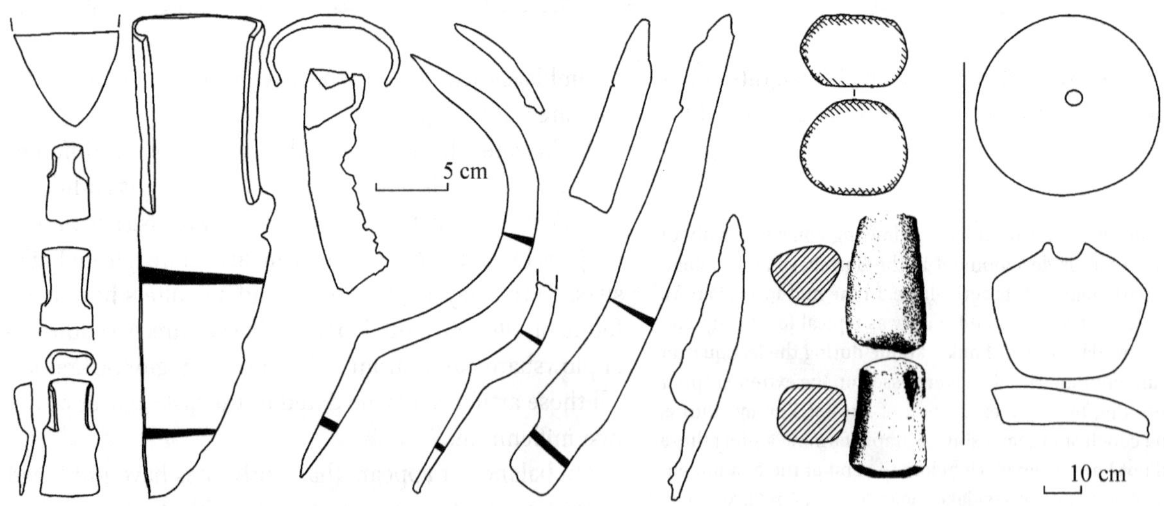

PLATE 6.3 Details of agricultural tools from Borshchevo sites

to further archaeological research. The find of asymmetrical plowshare is in this respect of significant interest. This large tool of a superior technological level was found in Krasne (Donetsk region, Ukraine).[12] Another such plowshare is known from the post-Saltiv (Slavic) occupation phase in the Mokhnach hillfort.[13] Together with that plowshare, two coulters were found, which are typical for the Saltiv culture. The evidence suggest therefore that the Slavs of the Romny culture learned about the superior forms of plowshares from the members of the Saltiv communities, and were ready to adopt those forms.

Coulters found on Saltiv sites have massive stems. Such coulters are rare in the Slavic cultures to the west and north-west from Khazaria (assemblages of the Volyntseve-Romny and Raiky cultures). In the southern parts of Eastern Europe, coulters found in Slavic sites are either stemmed or socketed. However, the former do not appear in the lands farther to the west.[14]

Similar observations may be made in relation to the iron reinforcements for spade blades: they have so far been found only on Saltiv sites. A fragment is known from the Zhivotinnoe hillfort attributed to the Slavs of the Don area.[15] Such tools were quite common at the end of the first millennium in the southern parts of Eastern Europe.[16] However, they do not appear either on Volyntseve-Romny or on Raiky sites. Given that the Zhivotinnoe hillfort has been attributed to the Borshchevo culture, one may surmise a certain influence from the Saltiv culture. However, iron reinforcements for spade blades tool did not become popular among the Severians. It was only later, during the Rus' age, that they appear in more Slavic sites.

Hoe-heads with a vertical disconnected socket appear in the southern parts of Eastern Europe as early as the 4th century (Sântana de Mureş-Chernyakhov culture). A few specimens appear also on sites attributed to the Pen'kovka culture, but they became popular only towards the end of the first millennium.[17] However, among the Slavs in the Don region, such hoe-heads are very rare.[18] A somewhat better presence has been documented on sites of the Volyntseve-Romny (about 20 specimens) and Raiky cultures (about 15 specimens). The number of specimens from Saltiv sites is far larger than that. As a matter of fact, Svetlana Pletneva was able to advance a typology for hoes on the basis of finds from catacomb cemeteries.[19] It is therefore likely that the presence of hoe-heads on sites of Slavic cultures is the result of the strong influence of the agricultural practiced in the Saltiv culture. A similar conclusion may be drawn on the basis of hoe-heads with horizontal socket, which are not known among the Slavs in the Don region. Such hoe-heads have been found on Slavic sites in Left-Bank Ukraine, particularly in the contact zone — in the hillforts at Vodiane (Kharkiv region, Ukraine) and Novotroits'ke (Sumy region, Ukraine). Hoes with a similar type of attachment are known from the hillforts in Monastyrek in the Middle Dnieper area (Cherkassy region, Ukraine) and Echimăuţi (Republic of Moldova), within the area of the Raiky culture.[20] The location of both sites is far removed from the contact zone under study in this book. Moreover, the tools found in Monastyrek and Echimăuţi are different from that found in Vodiane, albeit similar to that found in Novotroits'ke. The Vodiane specimen is in fact similar to hoe-heads from Saltiv sites, which suggests that the Slavs in Left-Bank Ukraine have adopted those types of hoes from the Saltiv culture. Those hoes then became very popular during the Rus' period. It is perhaps of interest that the hoard from Boiarka included hoe-heads with both types of attachment (three specimens for each type).[21]

There are also more types of tools for harvesting in the Saltiv than in the Slavic cultures: four types of sickles and two types of scythes. Sickles with a bent stem, which were common at the end of the first millennium, appear occasionally on sites of the Borshchevo culture (two specimens have been so far published, a whole tool, and a fragment). Much greater numbers of such sickles are known from sites of the Saltiv, Volyntseve-Romny, and Raiky cultures. On Slavic sites, several sickles have been found with attachments of different types, and with shapes and proportions that are typical for earlier periods (from the early Iron Age to the 7th century AD). However, all of them have been found on multilayered sites, and cannot therefore be dated with any precision to the last quarter of the first millennium AD. Shanked sickles such as found on Saltiv sites are known in

12 Mikheev 1985, fig. 23, *7*.

13 It is important to note that the plowshare was found together with the knife coulters lay at the base of the shaft. This suggest that the small hoard was buried, perhaps ritually, prior to construction of the bulwark during the Rus' period (Koloda 2002a).

14 Henning 1987, 53 fig. 21.

15 Vinnikov 2012, 130 fig. 8, *3*.

16 Henning 1987, 74 fig. 32.

17 Gorbanenko and Pashkevich 2010.

18 Only three specimens are known so far (Vinnikov 1995, fig. 11, *6*, *7*; Vinnikov 2012, fig. 8, *1*, *2*).

19 Pletneva 1989, 91–93.

20 Maksimov and Petrashenko 1988, fig. 74; Fedorov 1953, 120 fig. 51, *4*.

21 Gotun and Gorbanenko 2016.

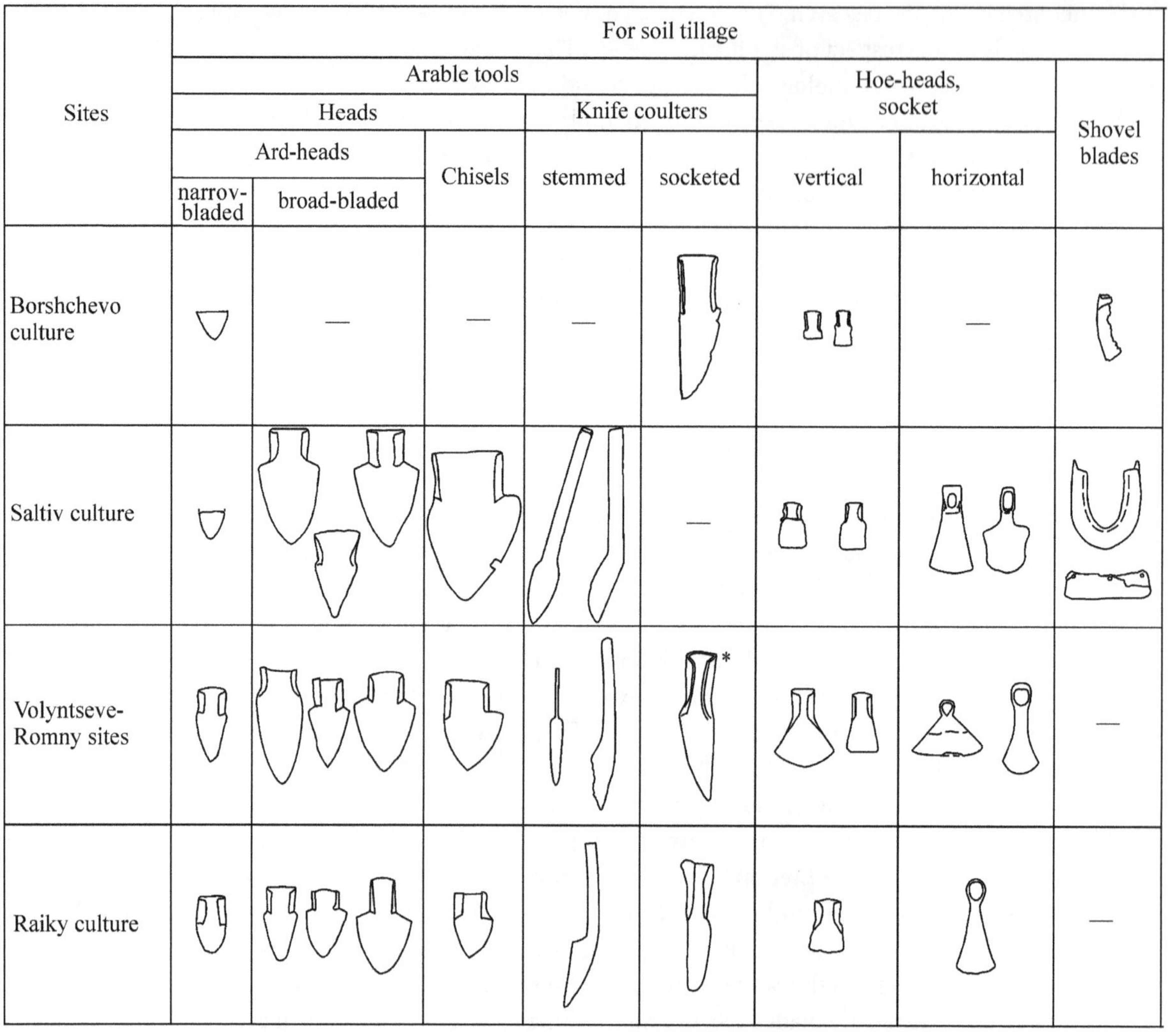

PLATE 6.4 Comparison of agricultural tool details dated to the last quarter of the first millennium. The symbol "+" indicates references in the scholarly literature.

the Donets region ever since the Scythian age.[22] They were common in all archaeological cultures dated from the end of the first millennium BC to the third quarter of the first millennium AD.[23] The shanked sickles of the Saltiv culture were a more advanced version of the type, the handle of which was attached to the shank by means of an elastic or metal clamp. This innovation may have responded to the need to move the otherwise cumbersome tool from one place to the other. If so, it is not at all surprising that the much more sedentary Slavs did not adopt this form. Sickles with a hooked type of attachment, which appear on Saltiv sites are isolated finds of archaic forms.

The same is true for the folding forms of sickles, which appear only on Saltiv sites. Given that such sickles have often been found in graves with weapons, they may have played different roles in different situations. When found on settlement sites, those were definitely agricultural tools. Since they were deposited in graves of warriors, one may think of them being close-combat weapons. This idea may even be substantiated by the fact that those were small tools, somewhat larger than knives with curved blades. However, given that the graves in which they were deposited were relatively rich, folding sickles were most likely tools meant to be a symbol of a certain social or property status (elder or landowner).[24]

Although scythes appear in all early medieval cultures of Eastern Europea that are compared in this chapter, only scythes of Mikheev's first group (scythes with bended heel and a shank on it) appear on Slavic sites. To this day, only one scythe of Mikheev's second group (scythes without heel, and hole near its end) is k known from the Borshchevo

22 Hrechko 2010, fig. 90, 3, 7.

23 Gorbanenko and Pashkevich 2010, figs. 7.2–7.7.

24 This matter definitely deserved a separate investigation.

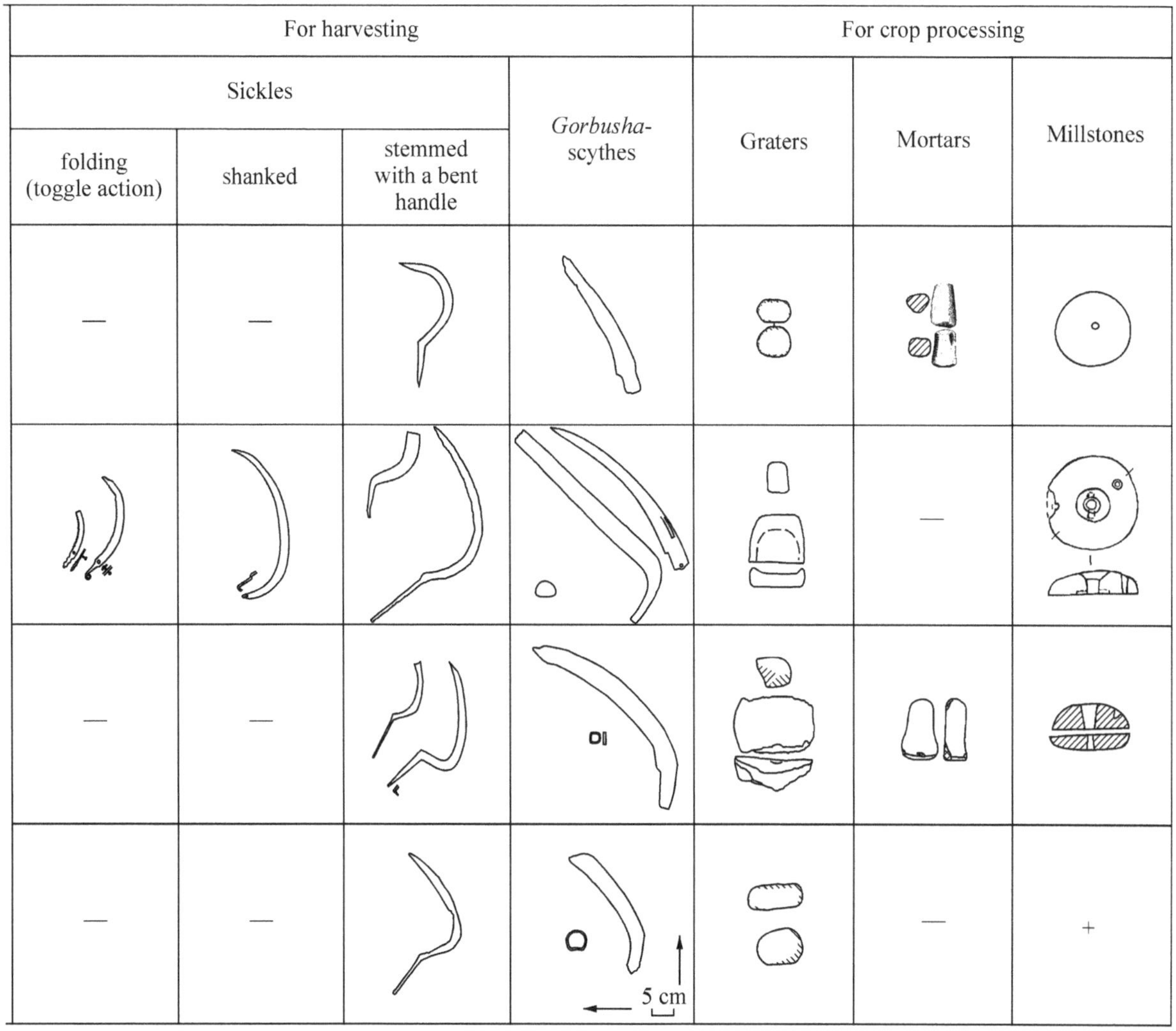

culture. Mikheev's first group was relatively common during the second quarter of the first millennium. In the Donets region, the type was documented on two sites of the Pen'kovka culture.[25] *Gorbusha*-type scythes appear at the end of the first millennium not only in the Saltiv, but also in the Volyntseve-Romny and Raiky cultures. There is no indication that the latter two adopted the *gorbusha*-type of scythe from the Saltiv culture, unlike the Slavs of the Don area, whom that type of scythe reached through a Saltiv intermediary. In short, the greater the number of scythe finds and the greater typological diversity in the Saltiv culture indicates betray their much greater role in animal husbandry for the population of Khazaria (see below).

When it comes to tools for processing crop, graters are the least informative for analysis. They are known from all cultures since Antiquity to the end of the first millennium AD, and they appear in the Saltiv, as well as in the neighboring Slavic cultures. To be sure, there are many more graters on sites attributed the Severians and to the Slavs of the Don area graters than on the sites of the Saltiv culture. On the latter, only a few graters have been found, which were used for grinding small amounts of groats or flour. Much more interesting is a stone pestle for a mortar, which was found in the post-Saltiv (Romny-age) occupation phase of the Mokhnach hillfort. To this day, this is a unique find in the whole part of Eastern Europe considered in this chapter. It may be that wooden pestles were used instead (they are known from ethnographic evidence), which left no traces in the record. By contrast, quern millstones were known to all the archaeological cultures at the end of the first millennium. When comparing numbers of finds, it becomes clear that the Slavs in the contact zone began using massive millstones under the influence of the Saltiv culture.

25 Gorbanenko and Pashkevich 2010, figs. 7.10–7.12.

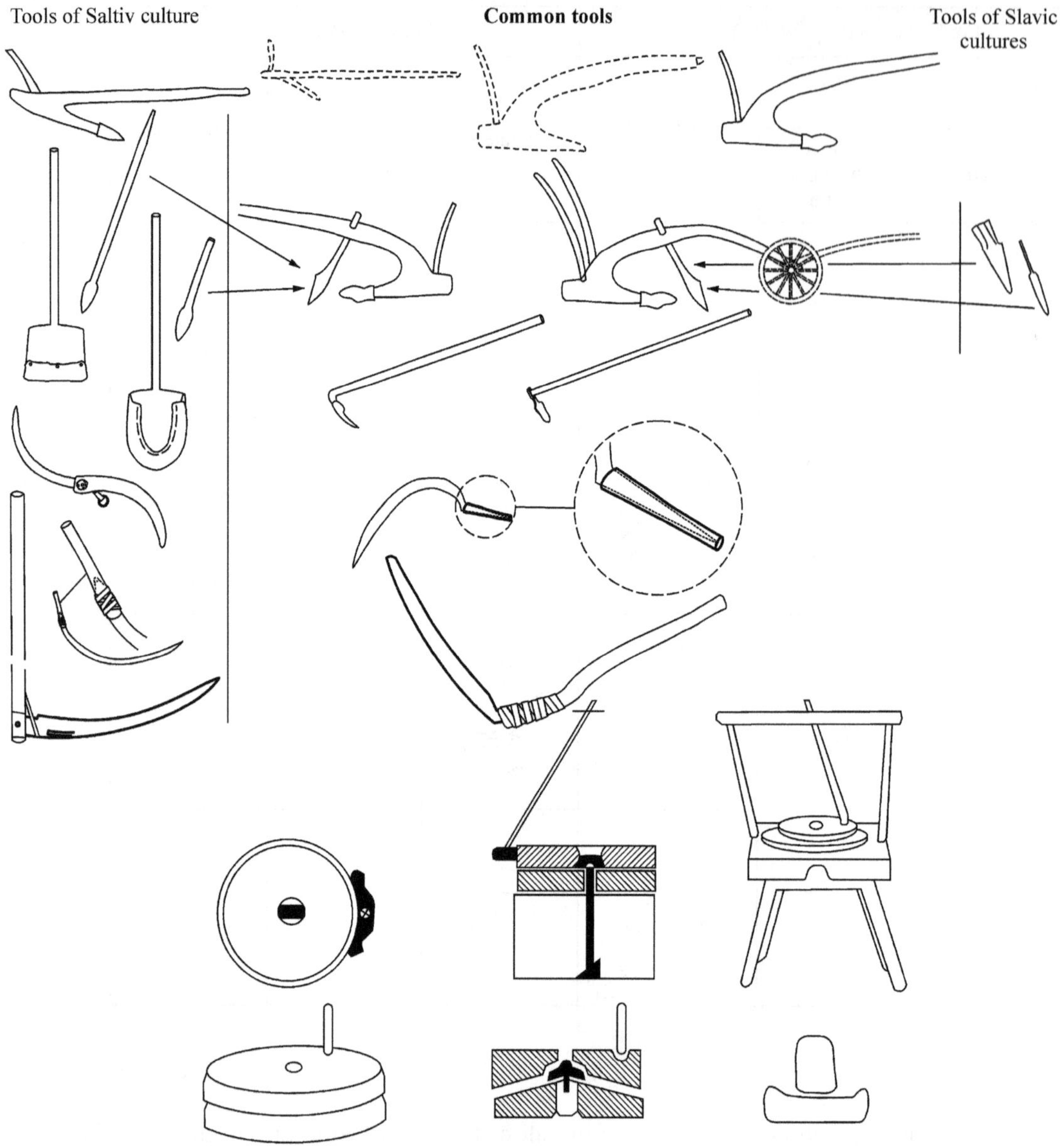

PLATE 6.5 Reconstruction of agricultural tools from sites of the Saltiv and Slavic cultures. Dashed line shows possible tools

The comparison based on details of the agricultural tools (see Plates 6.1–6.4) reveals both commonalities and differences. Plate 6.5 shows the most tools were common to both groups.

This implies not just a common tool kit, but also a comparable level of agricultural production. What about the tools that were unique either to one or to the other group? In the Donets region, narrow-bladed plowshares placed at an angle to the line of plowing appear only on Saltiv sites. Although this form of plowshare appears on Slavic sites outside of the contact zone, the lack of data for the area in question strongly suggests that in the contact zone Slavs had not put under cultivation land previously covered by forest. There were also some differences in the use of different other details of tools for primary soil tillage. The tool variety within the Saltiv culture concerns soil cultivation (in a broad sense), but presupposes a more developed horticulture. Moreover, such digging tools were also used for construction works. On the other hand, when it comes to tools for harvesting, the lack of folding sickles among the Slavs may indicates a greater degree of sedentary life than was the case for the Saltiv culture.

A meaningful comparison between archaeological cultures in Eastern Europe at the end of the first millennium is now possible because of a relatively large number of palaeobotanical studies published recently (Diagram 6.1).[26] In what follows, the basis of comparison is the PBS by

26 Gorbanenko 2012a; 2013; 2014a; 2014b; 2014c; 2014d; 2015a; 2015b; 2016; Gotun and Gorbanenko 2016; Chernenko et al. 2017.

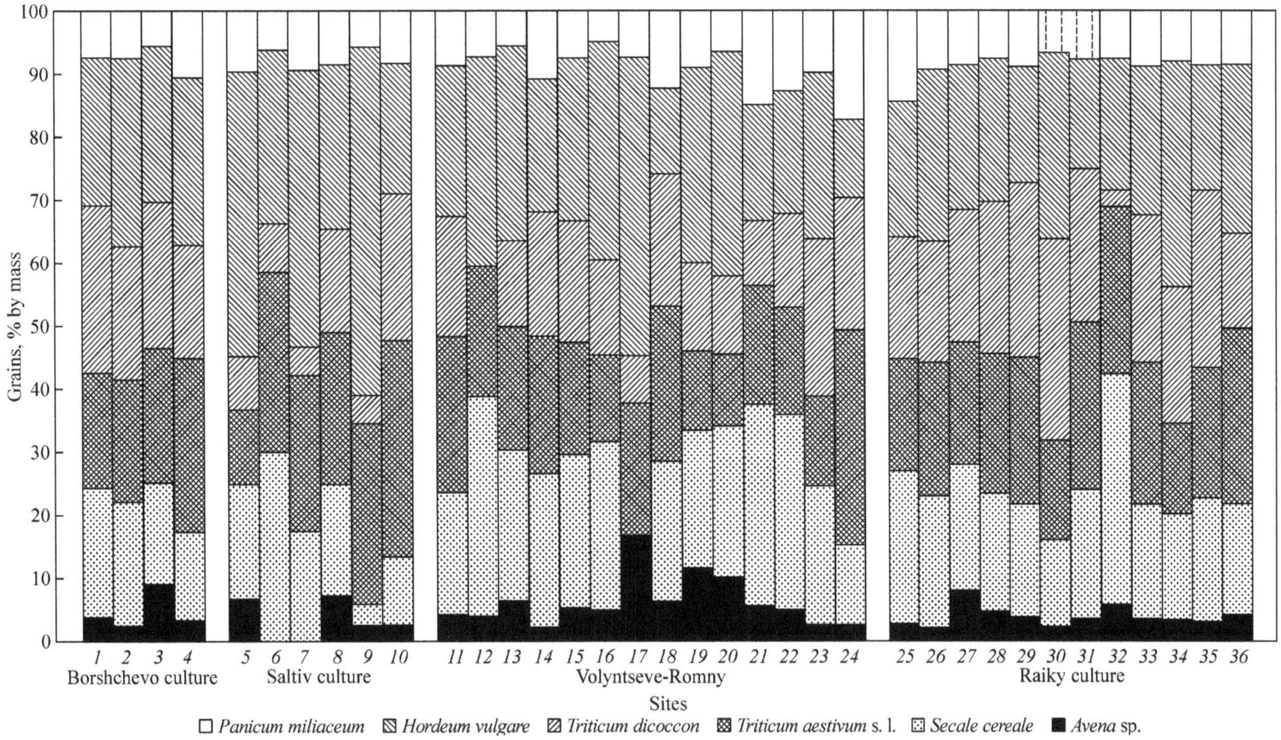

DIAGRAM 6.1 Comparison between palaeobotanical spectra of the last quarter of the first millennium. Sites: *Borschevo culture*: *1* — Belogor'e hillfort II; *2* — Large Borshchevo hillfort; *3* — Zhivotinnoe; *4* — Small Borshchevo hillfort and settlement; *Saltiv culture*: *5* — Verkhnyi Saltiv; *6* — Koropovi Khutory; *7* — Mokhnach, hillfort; *8* — Mokhnach P, settlement; *9* — P'iatnyts'ke I; *10* — Chuhuiv; *Volyntseve-Romny sites*: *11* — Bytytsa; *12* — Vodiane; *13* — Volyntseve; *14* — Glynsk; *15* — Gornal; *16* — Ltava; *17* — Mokhnach; *18* — Obukhiv II; *19* — Oleksandrivka I; *20* — Opishnia; *21* — Radychivs'ke 3; *22* — Sverdlovka I; *23* — Khodosivka, Kozakiv Yar; *24* — Shuklinka; *Prague culture*: *25* — Kodyn II; *26* — Kodyn I; *Raiky culture*: *27* — Hryhorivka; *28* — Luka-Raikovets'ka; *29* — Mala Snitynka; *30* — Pastyrs'ke; *31* — Peresopnytsa; *32* — Rashkiv I; *33* — Revne; *34* — Ridkivtsi; *35* — Sokil; *36* — Sokiltsi (dashed line shows possible indicators of proso millet)

weight, to the exclusion of imprints of millet on the bottoms of pots. The PBS of the cultures of the last quarter of the first millennium are indeed comparable. First, all have in common small proso millet, ranging from 5 to 10 percent. This must be interpreted as an indication of a high level of farming technology, far beyond slash-and-burn, which does not seem to have played any significant role in the economy at the end of the first millennium.[27] A large amount of the proso millet may be an indication of floodplain farming, as floodplain areas were cleared of weeds during floods. At the beginning of the vegetation period, proso millet is characterized by low-tallness, and therefore millet fields suffer from weed infestation.[28] Given the imprints of weed grains on pottery, it is therefore unlikely that floodplain farming played any essential role in comparable cultures, for weeds have been found on sites of Slavic cultures.[29]

The share of oats is also minor, even though they are present in small amounts in all palaeobotanical samples from archaeological cultures of the first millennium.[30] In a few cases, such as Rohalyk, oats occupy a more significant position in the PBS, perhaps because of its use in animal husbandry.[31]

Among all kinds of wheat, the hulled varieties are among the oldest plants that were cultivated, but naked varieties are far more productive.[32] Favoring naked over hulled wheat implies an improvement of cultivation technology.[33] The most reliable indications of wheat cultivation are those available at the moment for the Borshchevo culture.[34] Hulled and naked varieties of wheat represent about two fifths of the total number of

27 Tret'iakov 1932, 13–15. Proso millet was grown on a fairly wide expanse, not only on the cleared lands; an important fact for us is that low indicators of millet indicate a lack of slash and burn clearing as the main form of farming.

28 Hrihorovich 1933, pp. 7, 8; Yelagin 1955, p. 5; Lysov 1968, p. 8.

29 Kir'ianov 1967, 174; Pashkevich 1991c; Gorbanenko 2014d.

30 Pashkevich 1988, 171; Gorbanenko and Pashkevich 2010, 169–202.

31 Pashkevich and Gorbanenko 2004.

32 Brezhnev 1979, 213 and 214.

33 Lange 1975; Iazhdzhevskii 1988, 98–99.

34 Gorbanenko 2014d.

grains, with a slight predominance of the former variety over the latter. On sites attributed to other cultures, the share of wheat in the PBS varies between a fifth and a half. It is important to note that naked forms of wheat slightly outweigh the hulled forms in the Saltiv and the Volyntseve-Romny cultures.

The rye mass index is also interesting in that respect. Rye can grow on virtually any soils.[35] Scholars have also associated the preeminence of rye among all cultivated crops with the improvement of tools for soil tillage.[36] Indicators of rye also considerably vary on greatly, from null (Vodiane) to one third of the PBS. In general, rye and naked wheat together represent between one third and more than a half of the PBS. Those indicators are sufficiently large to support the idea of a high level of soil tillage.

Together with proso millet and hulled varieties of wheat, hulled barley is one of oldest cultivated cereals. Its share gradually decreased with the development of agriculture. However, if necessary (based on specific farming) it could be grown intentionally. For example, in Rohalyk, along with oats, barley was the basis of all palaeobotanical samples. This choice of cereal may indicate that plant cultivation was meant to meet the demands of animal husbandry, as both barley and oats were used for fodder. That barley has a higher share in the PBS of the Saltiv culture may indicate that animal husbandry, particularly horse breeding was more developed in the Saltiv, than in the neighboring Slavic cultures. The Slavic site on which hulled barley was as important as on Saltiv sites is the Vodiane hillfort.[37]

The greatest degree of similarity between PBS indicators is within the Borshchevo culture. This may well may because of the greater similarity of environmental conditions in which Borshchevo sites were placed, which led to very similar strategies in farming.

Similarities, however, may be found between Saltiv and Slavic cultures in terms of crop storing (plates 6.6–6.8). In both, large quantities of grain were likely kept for a long time in silos, mainly arranged outside dwellings. The walls of those pits may have been overlaid for a better protection of the grain against humidity. The nature of the coating cannot be determined archaeologically, but it may have been textile or timber. Silos have a well-defined, almost standardized form known both from archaeological studies, and from ethnographic data. The principle(s) of their usage was the same. A few variations have been noted on Saltiv sites, such as a barn build above two silos. Somewhat different methods storage were employed in the Kuznetsovskoe and Zhivotinnoe hillforts attributed to the Borshchevo culture. Square or rectangular pits were found on the site, which had postholes for supporting a covering made of wooden planks. Charred seeds of cereals were found in those pits. The excavators of those sites have interpreted the pits as silos.[38] However, judging from the small size of the features in question (2 to 4 sq m in area), their rectangular shape, and tight lining of their walls and floor with 3 to 4 cm — thick block, those were rather large underground chests. This may well be a feature characteristic for the Borshchevo culture, perhaps because the nature of the local soil made it impossible to dig pear-shaped silo, like those commonly found on Slavic sites.[39]

Minor amounts of grain were kept in small pits inside dwellings. Such pits appear on sites of all archaeological cultures at the end of the first millennium. To store food not only pits were used, but also containers, namely earthenware jars and large pots, which were found on sites of almost all comparable archaeological cultures. No study of the agriculture or the pottery of the Slavs in the Don area mentions anything about storage of crops in vessels, and no large vessels have been mentioned. However, Volodymyr Koloda, one of the authors of this book, was able to identify remains of jars among the Borshchevo pottery remains in the collection of Archaeological museum of the Voronezh State University. The capacity of the kitchen pots in the Borshchevo culture did not exceed 10 l. When grains are found scattered on the floor of the feature, this is interpreted as evidence of bags or sacks made of textile fabric or of leather, which served for keeping small amounts of grain at hand inside the house.

It appears therefore that the arable farming system was almost the same everywhere, no doubt, because the climatic conditions and the soils were alike. In both the Saltiv and the Slavic culture, fallow was predominant, an indication of either two- or three-field rotation. Differences existed of course, and they can be defined as follows. In the contact zone, probably, it was Saltiv communities that expanded the arable land by clearing the woods. There is no evidence of any clearing activity in the Romny culture of the contact zone. Once the land was

35 Brezhnev 1979, 276–277.

36 Iazhdzhevskii 1988, 88–89.

37 The palaeoclimatic study undertaken at Vodiane suggests that at the end of the first millennium, with increased aridity, barley may have been better adapted to the changing environmental conditions (Chendev and Koloda 2013), in the late 1st millennium AD a tendency towards aridization is traced, and a barley has better drought resistance as compared to other cereals.

38 Efimenko and Tret'iakov 1948, 103; Vinnikov 1995, 41–42.

39 Vinnikov 2014, 124–125.

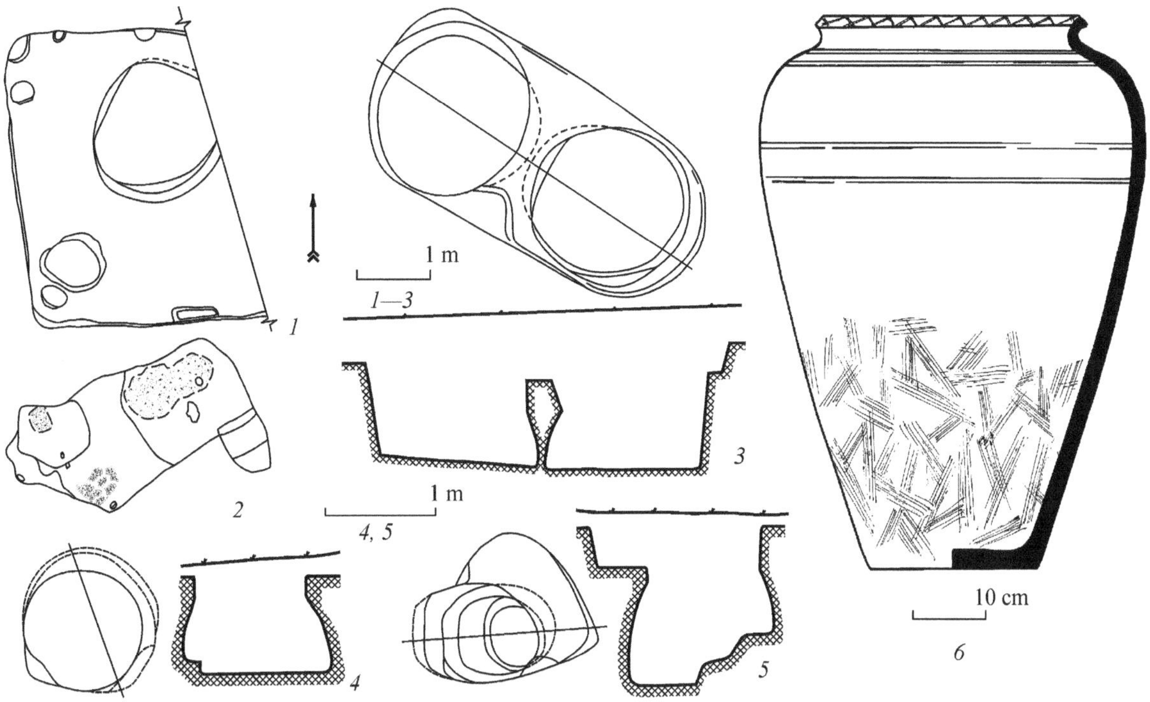

PLATE 6.6 Methods of grain storage on Saltiv sites: *1* — sunken-floored building with a grain pit; *2* — granary (without pits for storing grain); *3* — barn on two pits; *4* — cellar (with a pit for storing grain); *5* — grain pit; *6* — earthenware jar

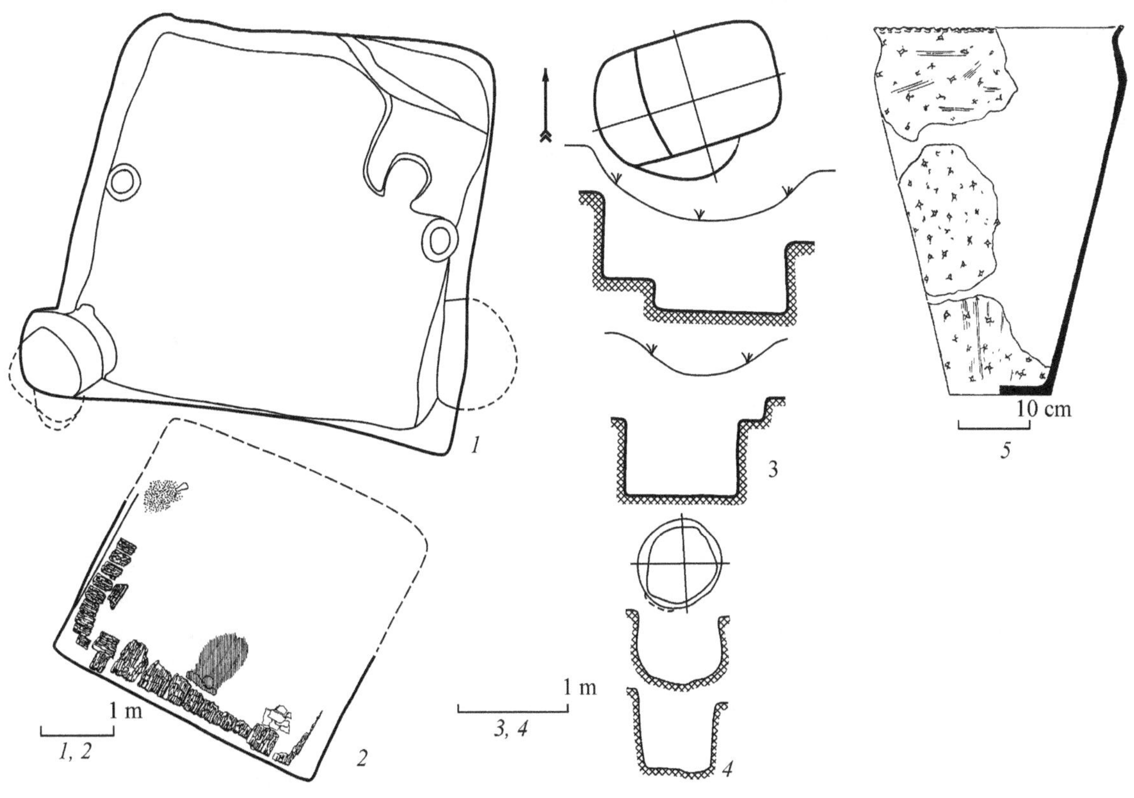

PLATE 6.7 Methods of grain among the Severians: *1* — sunken-floored building with a grain pit; *2* — above-ground outbuilding (without pits for storing grain); *3* — barn (with a pit for storing grain); *4* — grain pit; *5* — earthenware jar

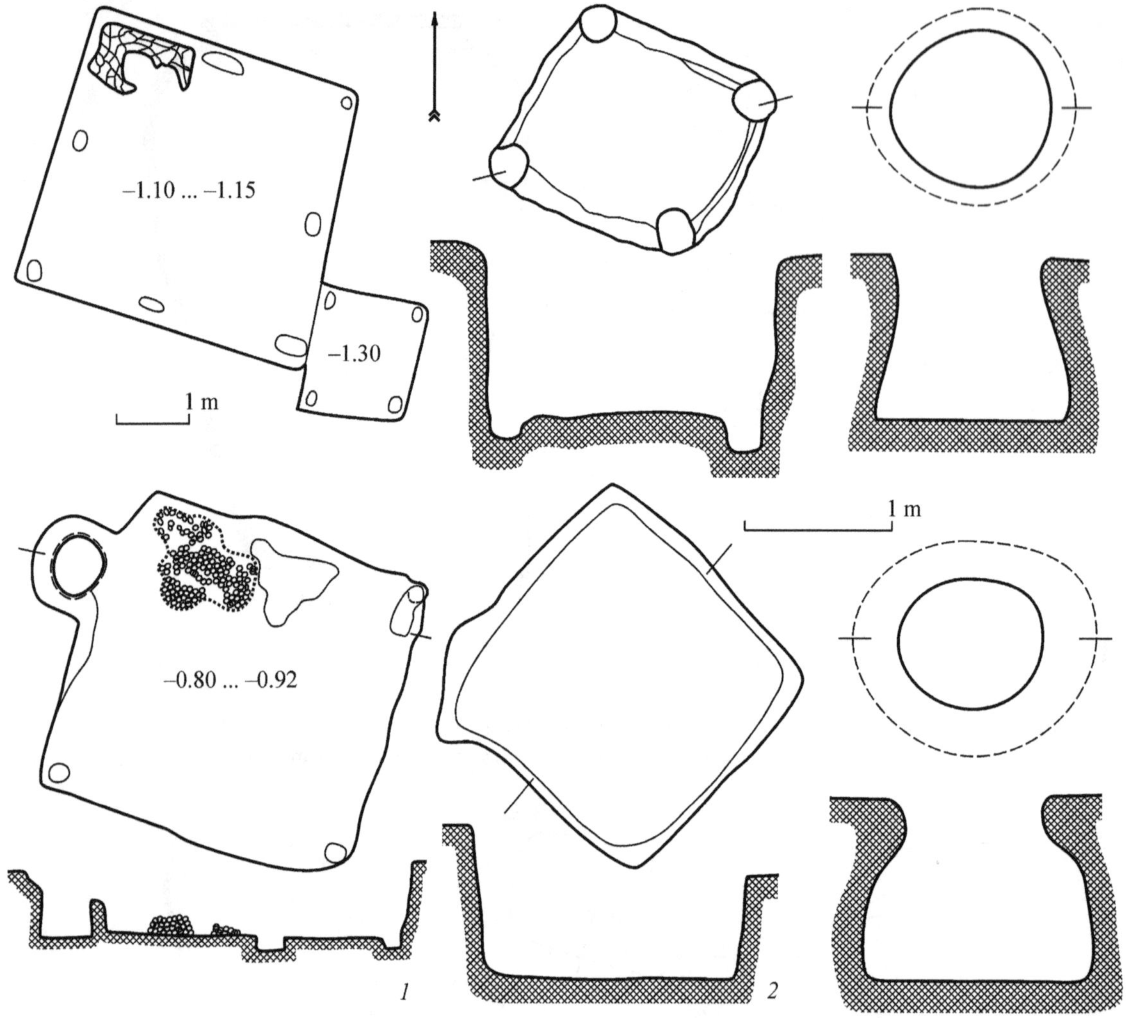

PLATE 6.8 Methods of grain storage on Borshchevo sites: *1* — dwellings with a place for food storage; *2* — rectangular pits; *3* — pear-shaped silos

gained for cultivation, the technologies involved in arable farming in the Saltiv and the Romny cultures were basically the same.

Recent discoveries at the border between the Kharkiv and Donetsk regions of Ukraine, have revealed that a part of the fields in the Saltiv culture may have been cultivated "by raid."[40] No less than eight assemblages of tools, including agricultural implements, have been found in a rather limited area near one of the southern spurs in the upper part of ta branched, wooded ravine that comes out from the right bank of the Donets and is called locally Gosudarev Iar.[41] All assemblages were found in shallow pits (between 50 and 70 cm deep). Despite the fact that those assemblages were not found during regular excavations, but accidentally by employees of the Sviati Hory National Park and by its visitors, it is remarkable that each one of them contains at least three groups of iron artifacts: farming tools (for soil tillage and crop storage), horse gear, and military equipment. Different other objects or groups of objects are sometimes added to this "core" — blacksmith, wood- and leather-working tools, as well as such artifacts as horse fetters and cattle bells. In a small trial excavation , quern stones and Saltiv pottery have been found.

A comprehensive study of the site by means of open-area excavation could of course provide ample evidence for interpretation. Nonetheless, a preliminary interpretation

40 Natural conditions in the valley of the river Donets are not unlike those of the forest-steppe area of the Saltiv culture, which may be regarded as the contact zone between the Slavs and the population of the Khazar Khaganate. This was in fact the ultimate reason for the archaeological exploration of the area.

41 Davydenko and Grib 2011; Koloda 2013. Archaeologists initially believed the site to be a cemetery, an assumption that later proved to be wrong, as no human bone remains and no traces of fire have been found in any of the eight features (Koloda 2014a; 2015, 116–119 with plate 1).

may be advanced even at this moment on the basis of the available r, materials. Each separate pit was a place where a single family stored its valuables — farming implements, craft tools, and household implements. Tools were buried in pits probably because out of the agricultural season, they were left where they would be required for work in the following year. They were buried in the upper part of the large-size, branched ravine because it was there that they could best be preserved. The blacksmith tools were there to repair agricultural implements or to make them on the spot, while the tools for leather- and woodworking were needed during seasonal agricultural labor. The presence of horse gear and weapons suggests that involved in the cultivation was not a sedentary, but probably a semi-nomadic (at least) population, which cultivated fields at a certain distance from its place of residence.

This is confirmed both by written and by ethnographic sources. According to the letter of the Khazar King Joseph, agricultural arable lands could be near settlement, as well as at a considerable distance of 20 *farsakhs*.[42] There is evidence prior of the Saltiv period of land in river valleys or in the steppe being under cultivation, at a considerable distance from settlements.[43] Arable fields scattered over a considerable distance were therefore cultivated "by raid," i.e., with only a temporary presence in the field. Taking into account the large distance, farmers could not reach their fields within one day. In the 16th century, both the Azov Tatars along the Donets, and the Crimean Tatars along the Oril and Samara rivers had such fields located at a considerable distance from their respective settlement.[44]

If our interpretation of the Gosudarev Iar will be proved to be correct, it will imply that an advanced form of arable farming, relying on quality implements, was associated not only with areas surrounding stationary, open settlements, but also with the campsites of the semi-nomadic population of the Saltiv culture.

3 Animal Husbandry

It is important to note that in animal bone assemblages from Slavic sites wild animals account for a fairly large amount, something that all scholars dealing with animal husbandry among the Slavs have noted. According to Veniamin Tsalkin, that bones of wild animals prevailed in assemblages from Romny-Borshchevo bespeaks the significant role of hunting.[45] Anatolii Vinnikov concluded that on Borshchevo sites, the share of wild animals in faunal assemblages varies between 45 to 60 percent.[46] At Monastyrek, a hillfort site in the Middle Dnieper area attributed to the Raiky culture, wild animals represent 36.6 percent of all animal bones.[47] The share of wild animals in faunal assemblages on Saltiv sites is considerably lower — between 2.4 to 9.5 percent. The only exception is Tsimliansk (assemblages from the 1987–1988 excavations), where the share reaches 41 percent (in older excavations, the share was supposedly close to a quarter). In other words, animal husbandry in the Saltiv culture occupied a more important place in providing food than in the neighboring Slavic cultures.

The animal husbandry of Saltiv population was also much more developed, as indicated both by the total percentage of meat of domestic animals in the diet, and by the greater number of stockbreeder tools. The reasons for the relatively important role of hunting among the Eastern Slavs at the end of the first millennium remain a matter of scholarly debate, and the matter deserved a much more detailed, separate study. For the moment, it is worth noting that hunting remained important on Slavic sites even in the contact zone. To be sure, the share of wild animals varies from one site to the other, and on some sites, reliable information is absent. Nonetheless, at least three reasons may be invoked for the situation in the contact zone. It has been notes that there are fewer Slavic than Saltiv sites in that zone, and each one of them is smaller in size than its Saltiv counterparts. It is therefore possible that at the beginning of their settlement in the Donets region, the Slavs put more emphasis on the main source of food — arable farming — and compensated the lack of protein of animal origin by means of hunting. Moreover, the sparse Slavic settlement of the contact zone enabled the long-term preservation of game, and made wild animals available for hunting. Although the archaeological evidence in that respect is missing, it is quite possible that hunting in the Slavic areas was largely based on trapping (nets, meshes, pits, loops, etc.), which greatly contributed to an increase in the amount of meat available. Moreover, the relations of the Slavs with the population of the neighboring Saltiv sites were influenced by the policy of the Khazar Khaganate. When those relations were peaceful, livestock products from the nearest Saltiv settlements were transferred to the Slavs on various conditions and by various

42 That is, between 120 and 140 km (Mikheev 1985, 43).

43 Kokovtsov 1932, 87–103; Mikheev 1985, 43; Pletneva 1986, 29; Dzhakson, Konovalova, and Podosinov 2009, 206.

44 Skrzhinskaia 1971, 150; Herberstein 1866, 153; Mikheev 1985, 43.

45 Tsalkin 1969, 92.

46 Vinnikov 1995, 45.

47 Belan 1978, 97. More recent studies of the animal bone assemblages from that site lowered the percentage to 29.3.

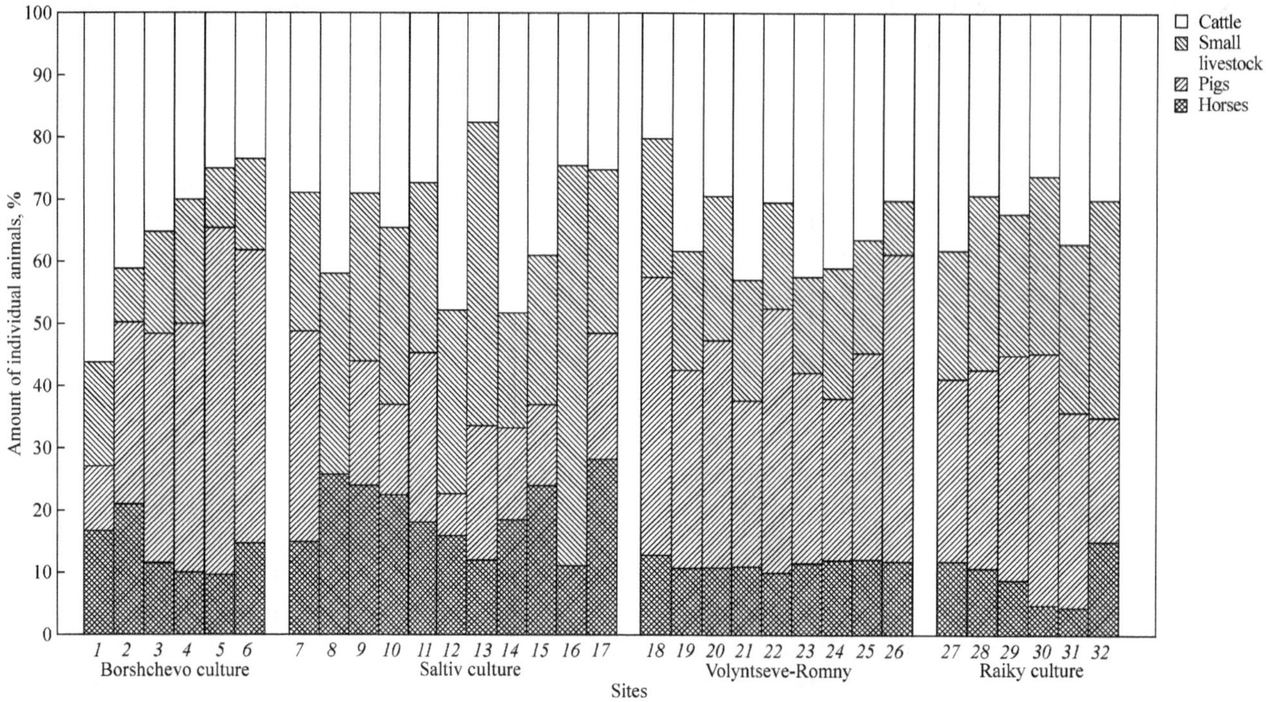

DIAGRAM 6.2 Comparison of domestic animals on sites of the last quarter of the first millennium. Sites: *Borschevo culture*: *1* — Arkhangelskie hillfort; *2* — Zhivotinnoe hillfort; *3* — Titchikha hillfort; *4* — Small Borshchevo hillfort; *5* — Large Borshchevo hillfort; *6* — Belogor'e I hillfort; *Saltiv culture*: *7* — Verkhnyi Saltiv; *8* — Nizhnegnilovskoe hillfort; *9* — Dmitrievka; *10* — Karnaukhivka; *11* — Koropovi Khutory; *12* — Lomakino; *13* — Mayaki hillfort; *14* — Mokhnach; *15* — Tsimliansk hillfort, excavations of the 1950s; *16* — Tsimliansk hillfort, excavations of 1987–1988; *17* — P'iatnyts'ke I; *Volyntseve-Romny sites*: *18* — Volyntseve; *19* — Glynsk; *20* — Ltava; *21* — Novotroits'ke; *22* — Obukhiv II; *23* — Opishnia; *24* — Petrivs'ke; *25* — Sverdlovka I; *26* — Khodosivka; *Raiky culture*: *27* — Monastyrek; *28* — Pastyrs'ke; *29* — Rashkiv I; *30* — Revne-Ripa; *31* — Revne-Tsaryna; *32* — Subotiv

means, thus delaying the development of a local system of cattle breeding. In case of conflict (usually in the form of raids from side of Khazaria), cattle was an immediate target, and the base of the local livestock decreased, which in turn forced the (surviving) Slavs to rely more on hunting.

Much more stable indicators of domestic animals have been recorded on sites of the Volyntseve-Romny culture, but it is worth considering animal husbandry in a broader perspective (Diagram 6.2). The most stable indicator of cattle is that of animal bone assemblages from Raiky sites — between a third and two fifths. Assemblages from Volyntseve-Romny sites are very similar, but on Volyntseve sites, the indicator sometimes drops to one fifth. By contrast, on Borshchevo sites cattle represents as little as a quarter (in 2 cases) and as much as 56.2 percent, but on most sites, the indicator varies between 30 and 41.2 percent. Assemblages in the Saltiv culture are a bit more stable than the figures for the Borshchevo culture — between 17.6 and 48.2 percent. It is important to note that, when it comes to cattle, unlike the situation for sheep or pig, there are no differences between Saltiv sites in the steppe and Saltiv sites in the forest-steppe zone.

The indicator for small livestock is quite stable on sits of the Volyntseve culture — one sixth to one fifth of all animal bones (again, the exception in Volyntseve, where the share drops to one tenth). Very similar is the situation on sites of the Raiky culture — one fifth to one third (the lowest value — 13.3 percent — is proportionally higher than that from Volyntseve sites). On Borshchevo sites, the indicator varies between one tenth and one fifth. However, on three out of six sites attributed to that culture, the indicator is as high as one sixth. In short, the situation in the Slavic culture concerning small livestock is relatively uniform.

A greater instability of the indicators pertaining to small livestock exists in the Saltiv culture. The lower figures match to the highest indicator for the Slavic sites — between a fifth and a quarter. However, there are also sites on which small livestock represents half of all animal bone assemblages, or is predominant. This suggests that animal husbandry strategies were specific to each natural zone (and probably site). Sheep were important in the steppe zone, where the Saltiv sites are located that have the largest indicators (Diagram 6.2).

The most stable indicator for pigs is from sites of the Volyntseve-Romny culture — between 26 and 49.4 percent, or between a quarter and a third. Very similar indicators have been recorded for Raiky sites — between 20 and 48.3 percent, or about a third. The indicator is less stable on the Slavic sites of the Don region — between a tenth and a little more than a half. There are considerable differences between sites. Equally unstable is the indicator for pig on sites of the Saltiv culture — from null to about a third.[48] In other words, pigs were raised by some Saltiv communities, but not by others. The variation is to be explained in terms of geography — sites in the steppe zone have either very low indicators, or no pig bones at all (e.g., the hillforts in Nizhnegnilovskoe and Tsimliansk, the latter according to the data collected in the 1980s). On the other sites (Karnaukhivka, Lomakino, but also Tsimliansk, according to the data collected in the 1950s), the indicator is never higher 14 percent (see Table 5.1). By contrast, on sites in the forest-steppe zone (Dmitrievka, Koropovi Khutory, Mayaki, Mokhnach, and P'iatnyts'ke I), where raising pigs is easier, the indicator is higher — between 15 and 34 percent (the highest indicator is at Verkhnyi Saltiv). One should not dismiss ethnic or religious factors as a possible explanation for pigs not being a main source of meat in Saltiv communities of the steppe zone.

On sites of the Saltiv culture, in general, horses represent between a sixth and a quarter of all animal bone assemblages (minimum of 11.1, maximum of 28.3 percent). In the Borshchevo culture, the indicator is one tenth in three cases, one sixth in two other cases, and one fifth in one case (variation from 9.6 to 21 percent). On sites of the Raiky culture, the indicator varies between 4.4 and 15 percent. On sites attributed to the Severians, horses represent about a tenth of all animal bones (between 10 and 12.8 percent). It appears therefore that horses played the least important role in the Raiky culture, with slightly better positions among Severiansand in the Borshchevo culture. Within the Saltiv culture, the indicator in two cases is similar to that recorded for sites attributed to the Siverians, while in all other cases it is similar or even higher than that of the Borshchevo sites. The comparison brings to the fore the importance of horse breeding in the life of Saltiv communities.

It is important to note in this respect that of all Saltiv sites, Verknyi Saltiv has produced indicators of different domestic animals that are closest to those of sites attributed to the Severians. Among the Slavic cultures, the indicators for Volyntseve-Romny and Raiky are the most stable, even though the composition of animal bone assemblages from sites attributed to those culture is sometimes different. The indicators for the Saltiv and Borshchevo cultures are comparatively less stable. Moreover, differences between them are quite significant. Within the Borshchevo culture, the most unstable factor is the share of pig bones, while in the Saltiv culture the most unstable are the indicators for small livestock and pigs. In other words, the animal husbandry of the Saltiv communities in the forest-steppe zone was quite similar to that of the neighboring Severians. In other words, the existing data show no specific influence of one culture over the other in terms of animal husbandry.

A few camel bones were found on Borshchevo sites (Titchikha and Large Borshchevo), much like in the Saltiv culture. This is direct evidence of the participation of the local Slavs, as well as the population of the Khaganate in the caravan trade along the Trans-Eurasian Silk Road.

Calculations of meat production (Diagram 6.3) bring out the differences in culinary preferences among the archaeological cultures considered in this book. They are in fact a reflection of the contrasts visible in the bone assemblages. Perhaps the most important thing to note is the absolute prevalence of beef in the diet, despite the fact that that prevalence is relative within each culture (one out of six sites in the Borshchevo culture; four out of 11 sites in the Saltiv culture; one out of nine sites in the Volyntseve-Romny culture; there are no sites with a prevalence of beef in the Raiky culture).

Tools for animal husbandry are isolated finds on the sites of the cultures being compared. Those tools are fewer in number than those for soil tillage. The most evident is the presence of spring shears for small livestock. Eight specimens from five sites are known from Saltiv sites in the Donets region, but only one specimen for the Borshchevo (from the Titchikha hillfort).[49] The specimen from the Bytytsa hillfort attributed to the Volyntseve-Romny culture is probably to be interpreted as the result of contacts with the Saltiv culture.[50] A single specimen is also known from the Raiky culture, and it was found in Ridkivtsi (Chernivtsi region, Ukraine).[51] That shears were more important in the Saltiv culture is therefore evident. Similarly, the only special butcher knives known from the region have been found on sites of the Saltiv culture (Plate 6.9).

48 The highest indicator has been recorded for Verkhnyi Saltiv.

49 Moskalenko 1965, fig. 5.

50 Gorbanenko 2012b, fig. 10.

51 Pyvovarov and Il'kiv 2014, fig. 5, 9.

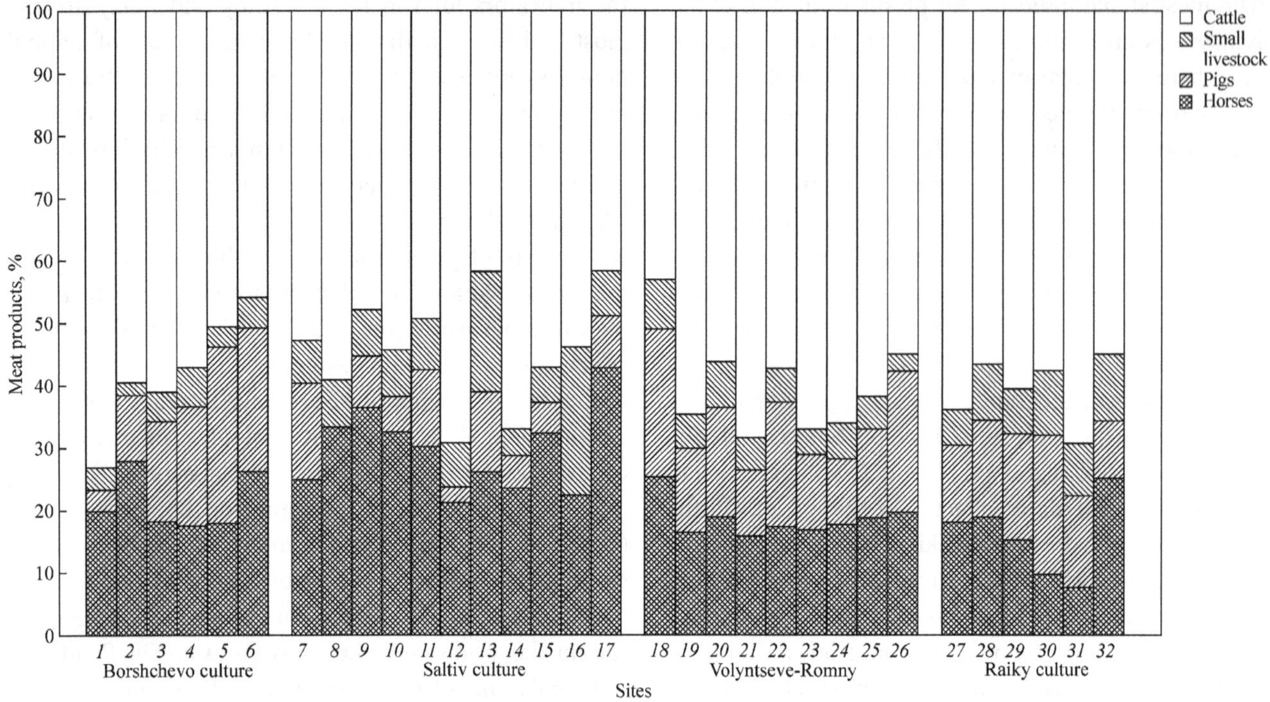

DIAGRAM 6.3 Comparison of meat products, obtained from livestock. Sites: *Borschevo culture*: *1* — Arkhangelskoe hillfort; *2* — Zhivotinnoe hillfort; *3* — Titchikha; *4* — Small Borshchevo hillfort; *5* — Large Borshchevo hillfort; *6* — Belogor'e I hillfort; *Saltiv culture*: *7* — Verkhnyi Saltiv; *8* — Nizhnegnilovskoe; *9* — Dmitrievka; *10* — Karnaukhivka; *11* — Koropovi Khutory; *12* — Lomakino; *13* — Mayaki hillfort; *14* — Mokhnach; *15* — Tsimliansk hillfort, excavations of the 1950s; *16* — Tsimliansk hillfort, excavations of 1987–1988; *17* — P'iatnyts'ke I; *Volyntseve-Romny sites*: *18* — Volyntseve; *19* — Glynsk; *20* — Ltava; *21* — Novotroits'ke hillfort; *22* — Obukhiv II; *23* — Opishnia; *24* — Petrivs'ke; *25* — Sverdlovka 1; *26* — Khodosivka; *Raiky culture*: *27* — Monastyrek; *28* — Pastyrs'ke; *29* — Rashkiv I; *30* — Revne-Ripa; *31* — Revne-Tsaryna; *32* — Subotiv

Saltiv sites have also produced large numbers of elements of the horse gear. Such finds are completely absent from the Borshchevo, and rare in the Volyntseve-Romny culture. It is important to note, however, that most details of the horse gear from the Saltiv culture have been found in burial, not settlement assemblages. No burial assemblages attributed to any of the Slavic cultures has produced such grave goods, but that is probably a matter of specific burial customs.

Shelters for animal are archaeologically documented for the Saltiv and Borshchevo cultures. For example, on the Titchikha hillfort of the Borshchevo culture, archaeologists found features with wide entrances indicated by postholes, which, according to the excavator, were meant for animals.[52] What make the Saltiv culture different in this respect, is the presence of stalls, which in two out of three cases were heated. Theoretically, it is not impossible that animals were kept inside pens within such Slavic hillforts, as Zhivotinnoe, Vodiane, or post-Saltiv Mokhnach. However, very little can be said about differences in animal husbandry on the basis of animal shelters and tools for animal breeding. The most important criterion for evaluation criterion remains the osteological material.

In conclusion, the comparative analysis reveals some differences in agriculture production. To be sure, the high technological level of soil tillage is about the same in all cultures being compared, even though there are many more agricultural in Khazaria, and of a greater variety. The same is true for soil tillage techniques. However, differences appear in harvesting, particularly in the absence or presence of folding sickles. Such tools are absent from all Slavic sites, both in the contact zone, and elsewhere. The influence of one culture over another in matters of farming techniques may be gauged by looking at the plowshares and coulters found in the post-Saltiv occupation phase of the Mokhnach hillfort, for which the closest analogies are in the Saltiv culture. The massive rotary millstones and the hoe-heads with horizontal sockets from Vodiane were most likely borrowed from the Saltiv milieu. It is equally possible that hoe-heads with a vertical socket were also adopted by the Slavs upon the influence of the Saltiv culture.

52 Moskalenko 1965, 61.

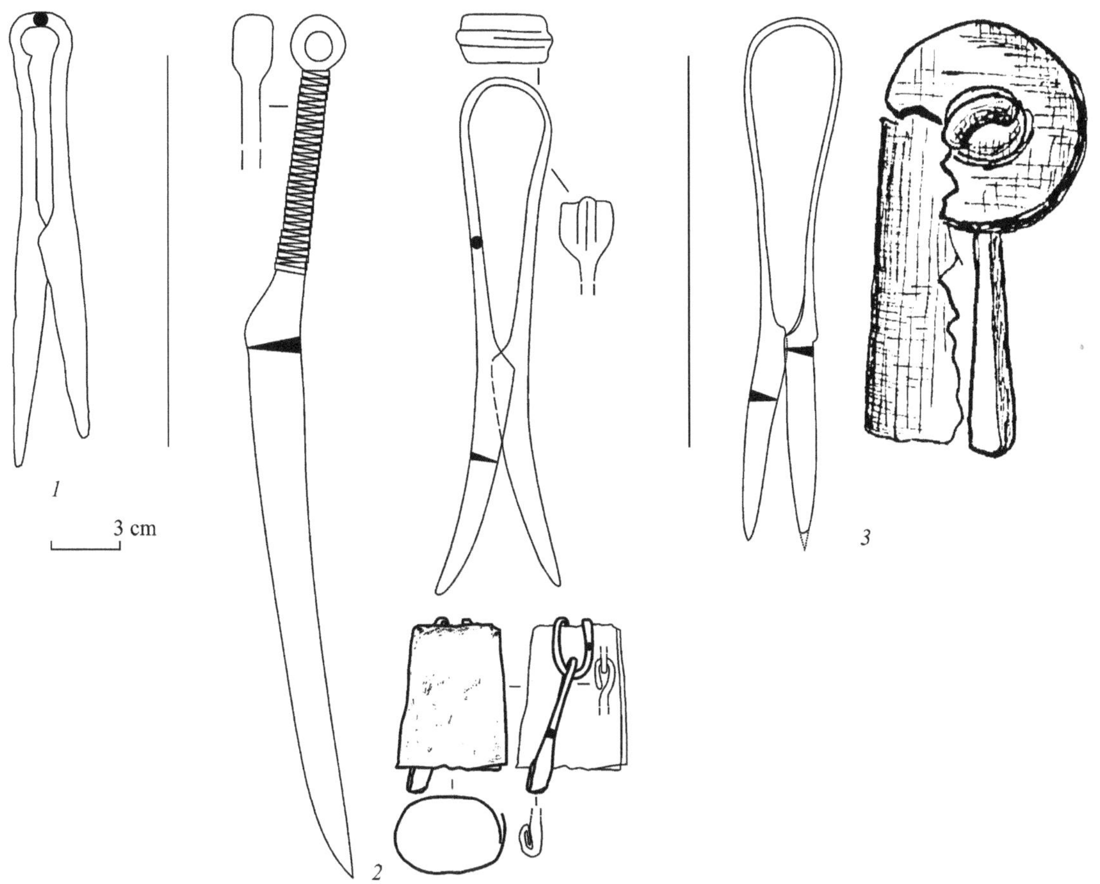

PLATE 6.9 Details of animal husbandry: 1 — *Borschevo culture*; 2 — *Saltiv culture*; 3 — *Volyntseve-Romny culture*

No signs of interaction between cultures, however, have been detected in animal husbandry. The differences noted above may therefore be explained in ethnic, social, or political terms, if they are not ultimately based on environmental conditions. The exploration of such explanatory possibilities is nonetheless the goal of a future study.

APPENDIX

The Anthracological Analysis of Samples from the P'iatnyts'ke I Settlement

Maryna Sergeyeva

No less than 68 charcoal samples have been submitted for analysis. Those samples were obtained by means of selective flotation of the soil from the pit 16. There were small pieces of charcoal in all samples.

Tree species have been identified after the characteristic features of their microstructure. The t results were then compared with data available data for species of wood.[1] In the case of charcoal satisfactorily preserved, wood can be defined by genus or, in some cases, by family.

The condition of preservation and the size of the charcoal pieces made it possible to identify species to the genus for 12 samples. Nine specimens were represented by deciduous trees of unidentified genus; among them two samples were scattered vascular deciduous species.

The results are as following:
ash (*Fraxinus* sp.) — 9;
oak (*Quercus* sp.) — 2;
rowan (*Sorbus* sp.) — 1;
deciduous *species* — 7;
scattered vascular deciduous species — 2.

The analyzed material includes exclusively deciduous species. Coniferous species are absent. All identified charcoal is of local provenance, as those wood genera are typical for the deciduous forests in the region in question. The results confirm the conclusion reached by palaeopedologists, namely that in ancient times the settlement in P'iatnyts'ke I was surrounded by deciduous forests.[2] The anthracological analysis has gone a step farther in identifying the species that made up those forests.

Because of its small size, the interpretation of the material is not possible. The value of those results is to expand the existing database on archaeological wood. For the moment, the study of fossil charcoal in Ukraine, particularly from settlements in Left-Bank Ukraine, is still in an incipient phase characterized primarily by the accumulation of data. A further examination of materials from different settlements may result in a sufficient amount of information to allow for the study of the palaeoecology and economic activity in the early Middle Ages. Moreover, this is the first targeted identification of tree species for sites of the Saltiv culture, at least in the Donets region.

Further anthracological studies will make it possible to draw conclusions that are more specific. Using sampling from different parts of any feature, with accurate positions, following studies will be able to move beyond identifying any tree species to interpretations of their use for different economic needs — fuel, construction, etc.

1 Sukachev 1940; Gammerman, Nikitin, and Nikolaeva 1946; Vikhrov 1959.

2 Matviishyna, Karmazynenko, Kvitkovs'kyi, and Zadverniuk 2012, 251–252.

 | DOI:10.1163/9789004429574_009

Bibliography

Afanas'ev 1987 — Afanas'ev, G. E. *Naselenie lesostepnoi zony basseina srednego Dona v VIII–IX vv. (alanskii variant saltovo-maiatskoi kul'tury)*. Moscow: Nauka.

Aksenov 2000 — Aksenov, V. S. "Pokhovannia z konem drugoi polovyny VIII–IX st. verkhn'oi techii r. Sivers'kyi Donets (za materialamy saltivs'kykh gruntovykh mogylnykiv." Ph.D. dissertation, Institute of archaeology of NAS of Ukraine. Kiev.

Aksenov 2001–2002 — Aksenov, N. V. "Kompleksy s konskimi nachel'nikami iz Verkhnesaltovskogo katakombnogo mogil'nika." *Stratum+*, no. 5: 270–81.

Aksenov 2005 — Aksenov, V. S. "Novi vypadovki znakhidky kompleksiv voiniv-vershynkiv saltivskogo chasu na Kharkivshchyni." *Arkheologichnyi litopys Livoberezhnoi Ukrainy*, nos. 1–2: 61–71.

Aksenov 2012 — Aksenov, V. S. "Katakombnye zakhoroneniia Verkhne-Saltovskogo arkkheologicheskogo kompleksa s T-obraznymi fibulami." *Saltovo-maiats'ka arkheologichna kul'tura: problemy ta doslidzhennia*, no. 2: 10–18.

Aksenov and Mikheev 2006 — Aksenov, V. S. and V. K. Mikheev. "Naselenie khazarskogo kaganata v pamiatnikah istorii i kul'tury. Sukhogomol'shanskii mogil'nik VIII–IX. vv." *Khazarskii al'manakh* 5: 4–306.

Aksenov, Evseenko, Riapolov 2016 — Aksenov, V. S., Evseenko, S. V., and Riapolov, V. M. "Bibliografiia nauchnykh i uchebno-metodicheskikh trudov A. A. Tortiki." *Khazarskii almanakh* 14: 419–437.

Aleksandrovskii 2002 — Aleksandrovskii, A. L. "Razvitie pochv Vostochnoi Evropy v golotsene." Ph.D. dissertation, Moscow State University. Moscow.

Aleksandrovskii and Aleksandrovskaia 2005 — Aleksandrovskii, A. L., and E. I. Aleksandrovskaia. *Evoliutsiia pochv i geograficheskaia sreda*. Moscow: Nauka.

Aleksandrovskii, Chendev, and Trubitsyn 2011 — Aleksandrovskii, A. L., Iu. G. Chendev, and M. A. Trubitsyn. "Paleopochvennye indikatory izmenchivosti ekologicheskikh uslovii Tsentra'lnoi lesostepi v pozdnem golotsene." *Izvestiia Rossiyskoi Akademii Nauk: Seriia geograficheskaia*, no. 6: 87–99.

Antipina 1997 — Antipina, E. E. "Metody rekonstruktsii osobennostei skotovodstva na iuge Vostochnoi Evropy v epokhu bronzy." *Rossiiskaia arkheologiia*, no. 3: 20–32.

Antipina 2005 — Antipina, E. E. "Miasnyie produkty v srednevekovom gorode — proizvodstvo ili potrebleniie?" In *Arkheologiia i estestvennonauchnye metody*. Edited by E. N. Chernykh and V. I. Zav'ialov, 159–180. Moscow: Iazyki slavianskoi kul'tury.

Antipina 2007 — Antipina, E. E. "Metody modelirovaniia otnositel'noi chislennosti domashnikh zhivotnykh v khoziaistve drevnikh poselenii: ot osteologicheskogo spektra k sostavu stada." *Materialy ta doslidzhennia z arkheologii Skhidnoi Evropy*, no. 7: 297–303.

Antipina 2008 — Antipina, E. E. "Sostav drevnego stada domashnikh zhivotnykh: logicheskie approksimatsii." *Opus: mezhdistsiplinarnye issledovaniia v arkheologii*, no. 6: 67–85.

Apareeva and Krasil'nikov 2011 — Apareeva, E. K., and K. I. Krasil'nikov. "Podvor'ia kak priznak khoziaistvennogo uklada naseleniia saltovo-maiatskoi kul'tury." In *Arkheologiia. Vid dzherel do rekonstruktsiy*. Edited by V. V. Otroshchenko, 186–92. Kiev: Instytut Arkheologii NAN Ukrainy.

Artamonov 1935 — Artamonov, M. I. *Srednevekovye poseleniia na nizhnem Donu (po materialam Severo-Kavkazskoi ekspeditsii)*. Leningrad: Gosudarstvennoe sotsial'no-ekonomicheskoe izdatel'stvo.

Artamonov 1936 — Artamonov, M. I. *Ocherki drevneishei istorii Khazar*. Leningrad Gosudarstvennyi sotsial'no-ekonomicheskii izdatel'stvo, 1936.

Artamonov 1958 — Artamonov, M. I. "Sarkel-Belaia Vezha." In *Trudy Volgo-Donskoi arkheologicheskoi ekspedicii*. Edited by Mikhail I. Artamonov, vol. 1, 7–84. Moscow/Leningrad: Izdatel'stvo Akademii Nauk SSSR.

Artamonov 1962 — Artamonov, M. I. *Istoriia khazar* Leningrad: Izdatel'stvo Gosudarstvennogo Ermitazha.

Babychev and Marynych 1993 — Babychev, F. S., and A. M. Marynych (eds.). *Geografichna entsiklopediia Ukrainy*, vol. 3. Kiev: Ukrains'ka entsyklopediia.

Bagalei 1905 — Bagalei, D. I. "Obiasnitelnyi tekst k arkheologicheskoi karte Kharkovskoi gubernii." In *Trudy XII Arkheologicheskago săezda v Kharkove 1902 g.*, vol. 1, 1–92. Moscow: Tovarishchestvo tipografii A. I. Mamontov.

Bakhteev 1956 — Bakhteev, F. Kh. "K istorii kultury iachmen'ia v SSSR." In *Materialy po istorii zemledeliia SSSR*, vol. 2, 210–215. Moscow: Izdatel'stvo Akademii Nauk SSSR.

Baran 2004 — Baran, Ia. V. *Slov'ians'ka obshchyna*. Kiev/Chernivtsi: Zelena Bukovyna.

Baranov 1990 — Baranov, I. A. *Tavrika v epokhu rannego srednevekov'ia (saltovo-maiatskaia kul'tura)*. Kiev: Naukova Dumka.

Barbarych 1977 — Barbarych, A. I. (ed.). *Geobotanichne raionuvannia Ukrains'koi RSR*. Kiev: Naukova dumka.

Belan 1978 — Belan, N. G. "Fauna gorodishcha Monastyrek na Srednem Dnepre." In *Ispol'zovanie metodov estestvennykh nauk v arkheologii. Sbornik nauchnykh trudov*. Edited by Vladimir F. Gening, 96–109. Kiev: Naukova dumka.

Berezovets 1952 — Berezovets, D. T. "Plug z Tokarivskogo torfovyshcha." *Arkheolohiia*, no. 7: 174–175.

Berezovets et al. 1986 — Berezovets', D. T., O. V. Parkhomenko, R. S. Orlov, A. T. Smilenko, and I. A. Baranov. "Rannesrednevekovye pamiatniki iuga Ukrainy." In *Arkheologiia Ukrainskoi SSR*. Edited by V. D. Baran, M. P. Kuchera, E. V. Maksimov et al., vol. 3, 212–48. Kiev: Naukova dumka.

Bezusko and Klymanov 1987 — Bezusko, L. G., and V. A. Klymanov. "Klimat i roslynnist rivnynnoi chastyny zakhidnoi URSR u piznio-pislialiodovykivvia." *Ukrainskyi botanichnyi zhurnal* 43, no. 3: 54–58.

Bezusko, Bezusko, and Grechishkina 2008 — Bezusko, L. G., A. G. Bezusko, and Iu. V. Grechishkina. "Palinologichni ta radiokhronologichni kharakterystyky vidkladiv pizn'ogo golotsenu rozrizu Lopans'ke (Ukraina, Kharkivska oblast)." In *Biostratygrafichni osnovy pobudovy stratygrafichnykh skhem fanerozoiu Ukrainy. Zbirnyk naukovykh prats' Instytutu geologichnykh nauk NAN Ukrainy*. Edited by P. F. Gozyk, 338–342. Kiev: Instytut geologichnykh nauk NAN Ukrainy.

Bobrinskii 1978 — Bobrinskii, A. A. *Goncharstvo Vostochnoi Evropy. Istochniki i metody izucheniia*. Moscow: Nauka.

Boldakov 1951 — Boldakov, E. V. *Zhizn' rek*. Moscow: Gostekhteorizdat.

Bondarchik 1979 — Bondarchik, V. K. (ed.). *Pamiatniki narodnoi arkhitektury i byta Belorussii. Metodicheskie posobie po vyiavleniiu i oiosianiiu*. Minsk: Polymia.

Borisenkov and Pasetskii 1983 — Borisenkov, E. P., and V. M. Pasetskii, *Ekstremal'nye prirodnye iavleniia v russkikh letopisiakh XI–XVII vv.* Leningrad: Gidrometeoizdat.

Borisov 1975 — Borisov, A. A. *Klimaty SSSR v proshlom, nastoiaschem i budushchem*. Leningrad: Izdatel'stvo Leningradskogo gosudarstvennogo universiteta.

Bregadze 1982 — Bregadze, N. *Ocherki po agroetnografii Gruzii*. Tbilisi: Metsniereba.

Brezhnev 1979 — Brezhnev, D. D. (ed.). *Kul'turnaia flora SSSR, 1. Pshenitsa*, Leningrad: Kolos.

Brockhaus and Efron 2002 — Brockhaus, F. A., and I. A. Efron, *Entsiklopedicheskii slovar'. Sovremennaia versiia*. Moscow: EKSMO (available online at http://gatchina3000.ru/brockhaus-and-efron-encyclopedic-dictionary/index.htm, visit of March 6, 2020).

Buniatian 1997 — Buniatian, K. P. "Skotarstvo ta sposib zhittia." *Arkheolohiia*, no. 3: 32–39.

Burroughs 2005 — Burroughs, W. J. *Climate Change in Prehistory. The End of the Reign of Chaos*. Cambridge/New York: Cambridge University Press.

Chendev 2008 — Chendev, Iu. G. *Evoliutsiia lesostepnykh pochv Srednerusskoi vozvyshennosti v golotsene*. Moscow: GEOS.

Chendev and Koloda 2012 — Chendev, Iu. G., and V. V. Koloda. "Prirodnaia sreda, pochvy i arkhitekturnye osobennosti gorodishcha Korobovy Khutora v Kharkovskoi obl. (problemy kompleksnogo izucheniia)." *Rossiiskaia arkheologiia*, no. 1: 112–121.

Chendev and Koloda 2013 — Chendev, Iu. G., and V. V. Koloda. "Pochvenno-arkheologicheskie issledovaniia na gorodishche Vodianoe: rekonstruktsiia prirodnoi sredy i khoziaistvennoi deiatel'nosti." *Drevnosti*: 225–236.

Chendev and Petin 2006 — Chendev, Iu. G., and A. N. Petin. *Estestvennye izmeneniia i tekhnogennaia transformatsiia komponentov okruzhaiushchei sredy staroosvoennykh regionov (na primere Belgorodskoi oblasti)*. Moscow: Izdatel'stvo Moskovskogo universiteta.

Chendev and Sarapulkin 2008 — Chendev, Iu. G., and V. A. Sarapulkin. "Pochvenno-arkheologicheskoe issledovanie pamiatnika 'Koltunovk'." In *Problemy prirodoispol'zovaniia i ekologicheskaia situatsiia v Evropeiskoi Rossii i sopredel'nykh stranakh. Materialy III Mezhdunarodnoi nauchnoi konferentsii (20–24 okt. 2008 g.)*, vol. 2, 139–143. Moscow/Belgorod: Politerra.

Chendev and Sarapulkin 2012 — Chendev, Iu. G. and V. A. Sarapulkin. "Opyt paleogeograficheskoi rekonstruktsii na primere issledovaniia pamiatnika saltovo-maiatskoi kul'tury v doline r. Tikhaia Sosna (Belgorodskaia oblast')." *Saltovo-maiats'ka arkheologichna kul'tura: problemy ta doslidzhennia*, no. 2: 96–99.

Chernenko et al. 2017 — Chernenko, O. E., R. A. Kravchenko, S. A. Gorbanenko, L. V. Gorobets', O. M. Kovalchuk, O. G. Seniuk and M. S. Sergeeva. "Sverdlovs'ke-1: biogospodars'ka diial'nist' meshkantsiv romens'kogo gorodyshcha na Chernigivshchyni." *Arkheologiia i davnia istoriia Ukrainy*, no. 1: 266–301.

Chernigova 2000 — Chernigova, N. V. "Babenko V. O. ta doslidzhennia arkheolohichnogo kompleksu v s. Verkhnyi Saltiv." *Arkheolohiia*, no. 4: 116–120.

Chisholm 1968 — Chisholm, M. *Rural Settlement and Land Use. An Essay in Location*. London: Hutchinson.

Davydenko and Grib 2011 — Davydenko, V. V. and V. K. Grib. "'Gosudarev iar' — novyi pamiatnik X–XI vv. v srednem techenii severskogo Dontsa (predvaritel'naia publikatsiia)." *Arkheologicheskii al'manakh* 25: 24–34.

Degtiareva 1981 — Degtiareva, G. V. *Pogoda, urozhai i kachestvo zerna iarovoi pshenitsy*. Leningrad: Gidrometeoizdat.

Dennell 1978 — Dennell, R. "Archaeobotany and early farming in Europe." *Archaeology* 31, no. 1: 8–13.

Dobiash-Rozhdestvenskaia 1936 — Dobiash-Rozhdestvenskaia, O. A. *Agrikultura v pamiatnikakh Zapadnogo srednevekovia: perevody i kommentarii pod redaktsiei*. Moscow: Izdatel'stvo Akademii Nauk SSSR.

Dokuchaev 1949 — Dokuchaev, V. V. *Sochineniia*, vol. 2. Moscow: Izdatel'stvo Akademii Nauk SSSR.

Dovzhenok 1952 — Dovzhenok, V. I. "Rozkopky bilia s. Volyntsevogo, Sums'koi oblasti." *Arkheologichni pam'iatky URSR* 3: 251–270.

Dovzhenok 1961 — Dovzhenok, V. I. *Zemlerobstvo drevn'oi Rusi.* Kiev: Vydanitstvo Akademii Nauk Ukrains'koi RSR.

Driesch 1976 — Driesch, A. von den. *A Guide to the Measurements to Animal Bones from Archaeological Sites, as Developed by the Institut für Paläoanatomie, Domestikationsforschung und Geschichte der Tiermedizin of the University of Munich.* Cambridge, Mass.: Harvard University Press.

Dubynin 1966 — Dubynin, A. F. "Ob ornamente gruzikov Troitskogo gorodishcha." *Sovetskaia arkheologiia,* no. 1: 269–274.

Dunlop 1954 — Dunlop, D. M. *The History of the Jewish Khazars.* Princeton: Princeton University Press.

Dzhakson, Konovalova, and Podosinov 2009 — Dzhakson, T. N., I. G. Konovalova, and A. V. Podosinov (eds.). *Drevniaia Rus v svete zarubezhnykh istochniko.: Khrestomatiia,* vol. 3. Moscow: Russkii fond sodeistviia obrazovaniiu i nauke.

Efimenko and Tret'iakov 1948 — Efimenko, P. P., and P. N. Tret'iakov. *Drevnerusskiie poseleniia na Donu.* Moscow/ Leningrad: Izdatel'stvo Akademii Nauk SSSR.

Efimova 1977 — Efimova, G. P. *Pochvennaia karta SSSR.* Moscow: Glavnoe upravlenie geodezii I kartografii.

Fedorov 1953 — Fedorov, G. B. "Gorodishche Ekimaucy." *Kratkie soobshcheniia Instituta Arkheologii AN SSSR* 50: 104–126.

Fernandez 2001 — Fernandez, H. "Ostéologie comparée des petits ruminants eurasiatiques sauvages et doméstiques (genres Rupicapra, Ovis, Capra et Capreolus): diagnose différentielle du squelette appendiculaire." Ph.D. dissertation, Université de Génève, 2 vols., Geneva.

Flerov 1994 — Flerov, V. S. "Pravoberezhnoe Cimlianskoe gorodishche v svete raskopok v 1987–1988, 1990 gg." *Materialy po arkheologii, istorii i etnografii Tavrii* 4: 441–516.

Frondzhulo 1968 — Frondzhulo, M. A. "O rannesrednevekovom remeslennom proizvodstve v iugo-vostochnom Krymu." In *Arkheologicheskie issledovaniia srednevekovogo Kryma.* Edited by O. I. Dombrovskii, 133–51. Kiev: Naukova dumka.

Fuks 1930 — Fuks, M. "Pro gorodishcha skitskoi doby na Kharkivshchyni." *Zapysky Vseukrainskoi arkheologichnoi komysii,* no. 1: 91–112.

Gadlo 1978 — Gadlo, A. V. "Kochevye Khazarskogo vremeni u stanicy Zaplavskoi na nizhnem Donu." *Problemy arkheologii,* no. 2: 118–125.

Gammerman, Nikitin, and Nikolaieva 1946 — Gammerman, A. F., A. A. Nikitin, and T. L. Nikolaeva. *Opredelitel' drevesin po mikroskopicheskim priznakam s a.'bomom mikrofotografii.* Moscow: Izdatel'stvo Akademii Nauk SSSR.

Gerenchuk 1964 — Gerenchuk, K. I. "Opyt klassifikatsii geograficheskikh landshaftov Ukrainskoi SSR i Moldavskoi SSR." *Voprosy regional'nogo landshaftovedeniia i geomorfologii SSSR,* no. 5: 5–12.

Golb and Pritsak 1982 — Golb, N. and O. Pritsak. *Khazarian Hebrew Documents of the Tenth Century.* Ithaca: Cornell University Press.

Golden 1983 — Golden, P. B. "Khazaria and Judaism." *Archivum Eurasiae Medii Aevi* 3: 127–56.

Golden 2005 — Golden, P. B. "Dostizheniia i perspektivy khazarskikh issledovanii." In *Khazary.* Edited by V. Ia. Petrukhin, W. Moskovich and A. M. Fedorchuk, 27–68. Jerusalem/ Moscow: Gesharim/Mosty Kul'tury.

Goleusov, Koloda, Lysets'kyi, Chendev 2002 — Goleusov, P. A., V. V. Koloda, F. N. Lysetskyi, and Iu. G. Chendev. "*Pochvy zemlianykh arkheologicheskikh pamiatnikov lesostepnoi zony i rekonstruktsiia po nim izmenenii prirodnoi sredy i pochvoobrazovaniia.*" *Vostochnoevropeiskii arkheologicheskii zhurnal,* no. 1, available online at http://archaeology.kiev.ua/journal /010102/goleusov_koloda_lisetsky_chendeyev.htm (visit of March 3, 2020).

Gorbanenko 2003–2004 — Gorbanenko, S. A. "Okrazhaiushchaia sred i slavianskie pamiatniki vtoroi poloviny I tys. n. e. (po materialam Levoberezh'ia Dnepra)." *Stratum+,* no. 5: 400–25.

Gorbanenko 2006 — Gorbanenko, S. A. "Zemlerobstvo slov'ian ostann'oi chverti I tys. n. e." *Arkheolohiia,* no. 3: 73–79.

Gorbanenko 2007 — Gorbanenko, S. A. *Zemlerobstvo i tvarinnytstvo slovian Livoberezhzhia Dnipra druhoi polovyny I tys. n. e.* Kiev: Akademperiodyka.

Gorbanenko 2008 — Gorbanenko, Sergei A. "Napivzemlianka no. 16 z Pastyrs'kogo gorodishcha." *Arkheologichnyi litopys Livoberezhnoi Ukrainy,* nos. 1–2: 50–53.

Gorbanenko 2009 — Gorbanenko, S. A. "Systemy zemlerobstva slov'ian kintsia I tys. do n. e.–I tys. n. e." *Muzeynyi visnyk,* no. 9: 85–103.

Gorbanenko 2012a — Gorbanenko, S. A. "Sel'skoe khoziaistvo do obrazovanie Kievskoi Rusi." In *Slaviane vostochnoi Evropy nakanune obrazovaniia drevnerusskogo gosudarstva. Materialy mezhdunarodnoi konferencii, posviashchennoi 110-letiiu so dnia rozhdeniia Ivana Ivanovicha Liapushkina (1902–1968), 3–5 dekabria 2012 g., Sankt-Peterburg.* Edited by O. A. Shcheglova and V. M. Goriunova, 106–09. St. Petersburg: SOLO.

Gorbanenko 2012b — Gorbanenko, S. A. "Sils'ke gospodarstvo zhyteliv Bytyts'kogo gorodyshcha." *Arkheolohiia,* no. 1: 98–114.

Gorbanenko 2013 — Gorbanenko, S. A. "Paleobotanichni doslidzhennia saltivs'kykh pam'iatok Lisostepy." *Saltovo-maiats'ka arkheologichna kul'tura: problemy ta doslidzhennia,* no. 3: 45–51, 148–50.

Gorbanenko 2014a — Gorbanenko, S. A. "Zernove gospodarstvo siverian." *Arkheolohiia,* no. 1: 113–123.

Gorbanenko 2014b — Gorbanenko, S. A. "Paleoetnobotanichni doslidzhennia slov'ians'kykh pam'iatok Bukovyny." *Arkheologichni studii*, no. 5: 200–211.

Gorbanenko 2014c — Gorbanenko, S. A. "Paleobotanichni materialy z NF IA NAN Ukrainy zi slov'ianskykh pam'iatok." *Arkheologiia i davnia istoriia Ukrainy*, no. 1: 181–187.

Gorbanenko 2014d — Gorbanenko, S. A. "Ril'nytstvo u nosiiv borshevs'koi kul'tury." *Arkheolohiia*, no. 2: 110–120.

Gorbanenko 2015a — Gorbanenko, S.i A. "Paleoetnobotanichni dani z poselennia Raykovets'koi kul'tury Mala Snitynka." *Arkheologiia*, no. 1: 109–116.

Gorbanenko 2015b — Gorbanenko, S. A. "Paleoetnobotanichni vyznachennia." *Arkheologichni doslidzhennia v Ukraini*: 244–248.

Gorbanenko 2016 — Gorbanenko, S. A. "Flotatsiia i promyvannia iak metody arkheologichnykh doslidzhen': realii i perspektyvy." *Arkheologiia i davnia istoriia Ukrainy*, no. 1: 137–144.

Gorbanenko and Koloda 2010 — Gorbanenko, S. A., Koloda, V. V. "Silske hospodarstvo meshkantsiv poselennia Verkhnyi Saltiv." *Arkheolohiia*, no. 1: 27–42.

Gorbanenko and Koloda 2013 — Gorbanenko, S. A., and V. V. Koloda. *Sil's'ke gospodarstvo na slov'iano-khozarskomu porubizhzhi*. Kiev: Instytut arkheologii NAN Ukrainy.

Gorbanenko and Pashkevich 2010 — Gorbanenko, S. A., and G. A. Pashkevich. *Zemlerobstvo davnikh Slov'ian (kinets' I tys. do n. e.–I tys. n. e.)*. Kiev: Akademperiodyka.

Gorbanenko, Koloda, and Pashkevich 2009 — Koloda, V. V., G. A. Pashkevich and S. A. Gorbanenko. "Zemlerobstvo zhyteliv gorodyshcha Mokhnach (chasiv saltivs'koi kul'tury)." *Arkheolohiia*, no. 2: 84–93.

Gorbanenko, Zhuravlev, and Pashkevich 2008 — Gorbanenko, S. A., O. P. Zhuravlev and G. A. Pashkevich. *Sil'ske gospodarstvo zhiteliv Pastyrs'kogo gorodyshcha*. Kiev: Akademperiodyka.

Gotun 1993 — Gotun, I. A. "Rekonstruktsii remisnychikh ta gospodarchikh budivel davn'orus'kogo poselennia Avtunichi." *Arkheolohiia*, no. 4: 59–71.

Gotun and Gorbanenko 2016 — Gotun, I. A. and S. A. Gorbanenko. "Zemlerobstvo zhyteliv slov'ians'kogo poselennia v uroch. Kozakiv iar u Khodosivtsi." *In Sclavenia terra* 1: 153–171.

Gromova 1950 — Gromova, V. I. *Opredelitel' mlekopitaiushchikh SSSR po kostiam skeleta*. Moscow: Izdatel'stvo Akademii nauk SSSR.

Gromova 1953 — Gromova, V. I. *Osteologicheskie otlichiia rodov* Capra *(kozly) i* Ovis *(barany). Rukovodstvo dlia opredeleniia iskopaemykh ostatkov*. Moscow: Izdatel'stvo Akademii nauk SSSR.

Gromova 1960 — Gromova, V. I. *Opredelitel po krupnym kostiam zapliusny*. Moscow: Izdatel'stvo Akademii nauk SSSR.

Guldenstedt 1804 — Guldenstedt, A. "Opisanie nekotorykh v Maloi Rossii upotrebitel'nykh plugov." *Tekhnologicheskii zhurnal* 1, no. 2: 3–31.

Guskov 2007 — Gus'kov, Anatolii. *Atlas peshchernykh gorodov Kryma. Puteshestvie k stroiteliam i obitateliam peshcher*. Kharkiv: Kursor.

Halstead, Collins, and Isaakidou 2002 — Halstead, P., P. Collins, and V. Isaakidou. "Sorting the sheep from the goats: morphological distinctions between the mandibles and mandibular teeth of adult *Ovis* and *Capra*." *Journal of Archaeological Sciences* 29: 545–553.

Henning 1987 — Henning, J. *Südosteuropa zwischen Antike und Mittelalter. Archäologische Beiträge zur Landwirtschaft des I. Jahrtausends u. Z.* Berlin: Akademie Verlag.

Herberstein 1866 — Herberstein, Sigmund Freiherr von. *Zapiski o Moskovii*. St. Petersburg: V Bezobrazov.

Hrechko 2010 — Hrechko, A. *Naselennia skifskogo chasu na Sivers'komy Dintsi*. Kiev: Instytut arkheologii NAN Ukrainy.

Hrihorovich 1933 — Hrihorovich, Iu. *Proso*. Kharkiv: Derzhhospvydav.

Ianushevich 1976 — Ianushevich, Z. V. *Kulturnye rasteniia Iugo-Zapada SSSR po paleobotanicheskim issledovaniiam*. Kishinev: Shtiintsa.

Ianushevich 1986 — Ianushevich, Z. V. *Kulturnye rasteniia severnogo Prichernomor'ia. Paleoetnobotanicheskie issledovaniia*, Kishinev: Shtiintsa.

Ianushevich and Markevich 1970 — Ianushevich, Z. V., and V. I. Markevich. "Arkheologicheskie nakhodki kul'turnykh zlakov na pervobytnykh poseleniiakh Pruto-Dnestrovskogo mezhdurech'ia." In *Introduktsiia kul'turnykh rastenii*. Edited by Z. D. Ianushevich, 83–110. Kishinev: Redaktsionoizdatel'stvo otdel Akademii nauk Moldavskoi SSR.

Iazhdzhevskii 1988 — Iazhdzhevskii, K. "O znachenii vozdelyvaemoi rzhi v kul'turakh zheleznogo veka v basseinakh Odry i Visly." In *Drevnosti slavian i Rusi*. Edited by B. O. Timoshchuk, 98–99. Moscow: Nauka.

Ielagin 1955 — Ielagin, I. I. "Proso na tselinnykh i staropakhotnykh zemliakh." *Znanie* 5: 9.

Jankuhn 1977 — Jankuhn, H. *Einführung in die Siedlungsarchäologie*. Berlin/New York: De Gruyter.

Jarman, Legge, and Charles 1972 — Jarman, H. N., A. J. Legge, and J. A. Charles. "Retrieval of plant remains from archaeological sites by froth flotation." *Papers in Economic Prehistory* 1: 39–48.

Kadeeev 1996 — Kadeeev, V. I. "Ukrainskii arkheolog, etnograf i kraeved V. A. Babenko." *Drevnosti. Kharkovskii istorikoarkheologicheskii ezhegodnik*: 144–146.

Karavaiko and Gorbanenko 2012 — Karavaiko, D. V., and S. A. Gorbanenko. *Gospodarstvo nosiiv iukhnivskoi kul'tury*. Kiev: Naukova dumka.

Khotinskii 1977 — Khotinskii, N. A. *Golotsen Severnoi Evrazii. Opyt transkontinental'noi korreliatsii etapov razvitiia rastitel'nosti i klimata*. Moscow: Nauka.

Kir'ianov 1959 — Kir'ianov, A. V. "Istoriia zemledeliia novgorodskoi zemli X–XV vv." In *Trudy novgorodskoi arkheologicheskoi ekspedicii*. Edited by Artemii V. Artsikhovskii and Boris A. Kolchin, 306–62. Moscow: Izdatel'stvo Akademii Nauk SSSR.

Kir'ianov 1967 — Kir'ianov, A. V. "Zemledelie vostochnogo slavianstva (VI–IX vv.)." In *Vozniknovenie i razvitie zemledeliia*. Edited by V. D. Blavatskii and A. V. Nikitin, 171–189. Moscow: Nauka.

Klimenko and Sleptsov 1999 — Klimenko, V. V., and A. M. Sleptsov. "Klimat i istoriia Rossii v IX–XVI vv." *Vestnik Moskovskogo energeticheskogo instituta*, no. 2: 85–93.

Kobylianskii 1989 — Kobylianskii, V. D. (ed.). *Kulturnaia flora SSSR* 2, 1. *Rozh*, Leningrad: Agropromizdat.

Kobylianskii and Luk'ianova 1990 — Kobylianskii, V. D., and M. V. Luk'ianova (eds.). *Kul'turnaia flora SSSR 2, 2: Iachmen'*. Leningrad: Agropromizdat.

Kokovtsov 1932 — Kokovcov, P. K. *Evreisko-khazarskaia perepiska v X veke*. Leningrad: Izdatel'stvo Akademii Nauk SSSR.

Koloda 2002a — Koloda, V. V. "Mokhnachanskaia nakhodka: klad ili zakladnaia zhertva?" In *Klady: sostav, khronologiia, interpretaciia. Materialy tematicheskoi nauchnoi konferencii*. Edited by Dmitrii G. Savinov, 112–16. St. Petersburg: Izdatel'stvo Sankt-Peterburgskogo Universiteta, Istoricheskii fakultet.

Koloda 2002b — Koloda, V. V. "Usad'ba srednevekovogo kuzneca na Mokhnachanskom gorodishche." *Khazarskii al'manakh* 1: 69–78.

Koloda 2002–2003 — Koloda, V. V. "Raboty 2003 g. na gorodishche Korobovy Khutora i v ego okruge." *Arkheologichni vidkryttia v Ukraini*: 167–169.

Koloda 2004 — Koloda, V. V. "Issledovaniia rannesrednevekovykh katakombnykh pogrebenii bliz s. Verkhnyi Saltov v 1996 g." *Khazarskii al'manakh* 3: 213–241.

Koloda 2005a — Koloda, V. V. "Postroiki dlia soderzhaniia skota u saltovskogo naseleniia Podonech'ia." In *Problemy doslidzhennia pam'iatok arkheologii skhidnoi Ukrainy. Materialy II-i Lugans'koi mizhnarodnoi istoryko-arkheologichnoi konferentsii, prysviachenoi 85-richchiu Lugan'skogo oblasnogo kraeznavchogo muzeiu*. Edited by V. V. Otroshchenko and S. S. Berezanskaia, 161–163. Luhans'k/Kiev: Shliakh.

Koloda 2005b — Koloda, V. V. Raboty na gorodische I selische Korobovy Khutora. *Arkheologichni doslidzhennia v Ukraini 2003–2004 rr.*: 171–173.

Koloda 2005–2007 — Koloda, V. V. "Arkheologicheskiie issledovaniia Kharkovskogo peduniversiteta v 2006 g." *Arkheolohichni doslidzhennia na Ukraini*: 218–222.

Koloda 2007 — Koloda, V. V. "Novye dannye o kontaktakh slavian Rusi i naseleniia Khazarskogo kaganata v svete khazaro-arabo-vizantiiskikh otnoshenii." In *Mezhdunarodnye otnosheniia v basseine Chernogo moria v drevnosti i srednie veka. Sbornik materialov XII Mezhdunarodnoi nauchnoi konferentsii*. Edited by V. P. Kopylov and A. N. Kovalenko, 92–95. Rostov-na-Donu: Izdatel'stvo Rostovskogo gosudarstvennogo pedagogicheskogo universiteta.

Koloda 2008 — Koloda, V. V. "Arkheologicheskii kompleks Korobovy Khutar: osnovnye itogi issledovanii." In *Problemy istorii i arkheologii Ukrainy. Materialy VI Mezhdunarodnoi nauchnoi konferencii, posviashchennoi 150-letiiu so dnia rozhdeniia akademika V. P. Buzeskula (g. Kharkov, 10–11 oktiabria 2008 g.)*. Edited by Vladimir I. Kadeev, 75–76. Kharkiv: Biznes Inform.

Koloda 2009a — Koloda, V. V. "Goncharnaia keramika volyntsevskoi kul'tury: istoricheskaia sud'ba rannikh severian i ikh mezhetnicheskie kontakty." *Divnogorskii sbornik*, no. 1: 164–78.

Koloda 2009b — Koloda, V. V. "Issledovanie remeslennogo tsentra saltovskoi kul'tury." *Arkheolohichni doslidzhennia v Ukraini*: 215–217.

Koloda 2010a — Koloda, V. V. "Selishche 'Mokhnach-P' v rannesrednevekovoi poselencheskoi strukture gorodishcha Mokhnach." In *Problemy istorii i arkheologii Ukrainy. Materialy VII Mezhdunarodnoi nauchnoi konferentsii (g. Kharkov, 28–29 okt. 2010 g.)*, 72. Kharkov: NTMT.

Koloda 2010b — Koloda, V. V. "Issledovanie remeslennogo tsentra saltovskoi kul'tury." *Arkheolohichni doslidzhennia v Ukraini*: 215–217.

Koloda 2012a — Koloda, V. V. "Eshche odna gruppa saltovskikh artefaktov iz Sukho Gomol'shi." *Saltovo-maiats'ka arkheologichna kul'tura: problemy ta doslidzhennia*, no. 2: 30–37, 108–113.

Koloda 2012b — Koloda, V. V. "Doslidzhennia 2012 r. na gorodishchi Mokhnach." *Arkheologichni doslidzhennia na Ukraini*: 338–339.

Koloda 2013 — Koloda, V. V. "Dva saltovskikh kompleksa iz Gosudareva Iara." *Saltovo-maiats'ka arkheologichna kul'tura: problemy ta doslidzhennia*, no. 3: 73–81, 161–168.

Koloda 2014a — Koloda, V. V. "O khoziaistvennykh iamakh so standartizirovannymi naborami predmetov na saltovskikh poseleniiakh." In *Problemy istorii i arkheologii Ukrainy. Materialy IX Mezhdunarodnoi nauchnoi konferentsii (Kharkov, 30–31 oktiabria 2014 g.)*, 54–55. Kharkov: NTMT.

Koloda 2014b — Koloda, V. V. "'Skarb' zemlerobskykh znariad z gorodishcha Koropovi Khutory." *Starozhytnosti livoberezhnogo Podniprov'ia*: 40–49.

Koloda 2014c — Koloda, V. V. "Specializirovannye remeslennye instrumenty saltovskogo perioda na gorodishche Mokhnach."

In *Arkheologiia: mozhlivosti rekonstruktsii*. Edited by S. A. Skorii, 68–74. Kiev: Instytut arkheologii NAN Ukrainy.

Koloda 2015 — Koloda, V. V. "Unikal'nyi zhilishchno-khoziaistvennyi kompleks saltovskoi kul'tury na selishche Mokhnach P." *Khazarskii al'manakh* 13: 111–29.

Koloda 2017 — Koloda, V. V. "Specializovani znariaddia iak vidobrazhennia rivnia remisnychogo vyrobnytsva na selyshchi 'Mokhnach P' u saltivskyy chas." *Arkheologiia i davnia istoriia Ukrainy*, no. 1: 40–49.

Koloda and Gorbanenko 2009 — Koloda, V. V. and S. A. Gorbanenko. "Zemlerobstvo saltivskoi kul'tury (za materialamy Mokhnachans'kogo gorodyshcha)." *Vita antiqua* 7–8: 261–80.

Koloda and Gorbanenko 2010 — Koloda, V. V. and S. A. Gorbanenko. *Sel'skoe khoziaistvo nositelei saltovskoi kul'tury v lesostepnoi zone*. Kiev: Institut arkheologii NAN Ukrainy.

Koloda and Gorbanenko 2012 — Koloda, V. V. and Sergei A. Gorbanenko. "Zemledelie na poselenii Korobovy Khutora v saltovskie vremia." In *Stepi Evropy v epokhu srednevekov'ia. 9. Khazarskoe vrem'i*. Edited by A. V. Evgelevskii, 119–134. Donetsk: Izdatel'stvo Donetskogo nacional'nogo universiteta.

Koloda and Koloda 2001 — Koloda, Vladimir V. and T. A. Koloda. "Starozhytnosti s. Mokhnach ta iogo okrugi." *Arkheologichnyy litopys Livoberezhnoi Ukrainy*, no. 2: 42–45.

Koloda and Koloda 2005–2009 — Koloda, V. V. and T. A. Koloda. "Keramika ranneslavianskogo gorodishcha Vodianoe na Khar'kovshchine." *Stratum+*, no. 5: 249–64.

Koloda and Kroitor 2015 — Koloda, V. V. and R. V. Kroitor. "Zhivotnovodstvo naseleniia saltovskoi kul'tury v lesostepnoi chasti Severskogo Donca (Ukraina)." *Revista Arheologică* 11, nos. 1–2: 273–93.

Koloda and Kroitor 2016 — Koloda, Vladimir V. and Roman V. Kroitor. "Zhivotnovodstvo naseleniia saltovskoi kul'tury v lesostepnoi chasti Severskogo Donca (Ukraina). Chast' 2." *Revista Arheologică* 12, 1–2: 296–303.

Koloda and Kroitor 2017 — Koloda, V. V. and Roman V. Kroitor. "Zhivotnovodstvo naseleniia saltovskoi kul'tury v lesostepnoi chasti Severskogo Donca (Ukraina)." *Revista Arheologică* 13, nos. 1–2: 220–34.

Koloda and Kvitkovskii 2008 — Koloda, V. V., and V. I. Kvitkovskii. "Arkheologicheskie issledovaniia KhNPU na territorii Kharkovskoi obl. v 2008 g." *Arkheolohichni doslidzhennia v Ukraini*: 146–149.

Koloda, Kroitor, and Gorbanenko 2013 — Koloda, V. V., R. V. Kroitor and S. A. Gorbanenko. "Novi dani pro tvarynnytstvo rann'oseredn'ovichnogo naselennia Verkhn'ogo Saltova." *Arkheolohiia*, no. 1: 129–138.

Koloda, Lysets'kyi, and Chendev 2002–2003 — Koloda, V. V., F. N. Lysets'kyi, and Iu. G. Chendev. "Pochvy arkheologicheskikh pamiatnikov lesostepnoi zony i rekonstruktsiia po nim izmenenii prirodnoi sredy i pochvoobrazovaniia." *Arkheolohichni doslidzhennia v Ukraini*: 163–167.

Koloda, Chendev, Borbukova, and Dudin 2014 — Koloda, V. V., Iu. G. Chendev, D. A. Borbukova, and D. I. Dudin. "Evoliutsiia pochv i prirodnoi sredy na iuzhnom uchastke vostochnoevropeiskoi lesostepi (na primere issledovaniia gorodishch Kharkovkoi oblasti)." *Povolzhskaia arkheologiia*, no. 1: 247–267.

Korobushkina 1979 — Korobushkina, T. N. *Zemledelie na territorii Belorussii v X–XIII vv.* Minsk: Nauka i tekhnika.

Kovalenko 1999 — Kovalenko, V. P. "Novi doslidzhennia Shestovyts'kogo arkheologichnogo kompleksu." *Arkheologichnyi litopys Livoberezhnoi Ukrainy*, no. 1: 33–43.

Kovalevskii and Gorbanenko 2014 — Kovalevskii, V. N. and S. A. Gorbanenko. "Znakhidka zaliznogo rala v rayoni Zhyvotynnogo gorodyshcha." *Arkheolohiia*, no. 3: 104–07.

Kozak 2008 — Kozak, D. N. *Venedy*. Kiev: Instytut arkheologii NAN Ukrainy.

Kozak and Pashkevich 1985 — Kozak, D. N., and G. O. Pashkevich. "Zemlerobstvo plemen Verkhn'ogo Podnistrov'ia i Zakhidnogo Pobuzhzh'ia na rubezhi ta v pershykh vikakh nashoi ery." *Arkheolohiia* 50: 18–27.

Kozak, Prishchepa, and Shkoropad 2004 — Kozak, D. N., B. A. Prishchepa, and V. V. Shkoropad. *Davni zemleroby Volyni (pamiatky arkheologii na Khrinnytskomu vodoimyshchi)*. Kiev: Instytut arkheologii NAN Ukrainy.

Krasil'nikov 1981 — Krasil'nikov, K. I. "Voznikovenie osedlosti u prabolgar Srednedonech'ia." *Sovetskaia Arkheologiia*, no. 4: 110–125.

Krasil'nikov 2011 — Krasil'nikov, K. I. Review of *Sel'skoe khoziaistvo nositelei saltovskoi kul'tury v lesostepnoi zone*, by V. V. Koloda and S. A. Gorbanenko (Kiev, 2010)]. *Arkheolohiia*, no. 3: 135–37.

Krasil'nikov 2012 — Krasil'nikov, K. I. "Khoziaistvennye struktury naseleniia saltovo-maiackoi kultury (SMK) (po materialom stepnogo Podoncov'ia)." *Divnogorskii sbornik* 3: 177–92.

Krasil'nikova 2005 — Krasil'nikova, L. I. "Budivli saltovo-maiatskoi kul'tury stepovogo Naddintsivia." Ph.D. dissertation, Institute of archaeology of NAS of Ukraine. Kiev.

Krasnov 1967 — Krasnov Iu. A. "O sisteme i tekhnike rannego zemledeliia v lesnoi polos'e Vostochnoi Evropy." *Sovetskaya arkheologiia*, no. 1: 3–21.

Krasnov 1969 — Krasnov, Iu. A. "Nekotoryie itogi i zadachi izucheniia istorii zemledeliia v sovetskoi arkheologii." *Kratkiie soobshcheniia Instituta arkheologii AN SSSR* 118: 58–68.

Krasnov 1971 — Krasnov Iu. A. *Rannee zemledelie i zhivotnovodstvo v lesnoi polose Vostochnoi Evropy (II tysiacheletiia do n. e.-pervaia polovina I tysiacheletiia n. e.)*. Moscow: Nauka.

Krasnov 1987 — Krasnov, I. A. *Drevnie i srednevekovye pakhotnye orudiia Vostochnoi Evropy*. Moscow: Nauka.

Kravchenko and Davydenko 2001 — Kravchenko, E. E. and V. V. Davydenko. "Sidorovskoe gorodishche" In *Stepi Evropy v epokhu srednevekov'ia. Khazarskoe vremia*. Edited by A. V. Evgelevskii, 233–302. Donetsk: 2001.

Kravchenko and Pashkevich 1985 — Kravchenko, N. M. and G. A. Pashkevich. "Nekotorye problemy paleobotanicheskikh issledovanii (po materialam Obukhovskoi territorial'noi gruppy pamiatnikov I tys. n. e.)." In *Arkheologiia i metody istoricheskikh rekonstrukcii. Sbornik nauchnykh trudov*. Edited by Vladimir F. Gening, 177–90. Kiev: Naukova dumka.

Kroitor 2013 — Arkheozoologichni kompleksy iz saltivs'kikh I romens'kikh pamiatok baseinu Sivers'kogo Dintsia. In Gorbanenko, S. A. and V. V. Koloda. *Sil's'ke gospodarstvo na sloviano-khazars'komu porubizhzhi*: 181–199, Kiev.

Kroitor and Koloda 2015 — Kroitor, R. V., Koloda, V. V. *Zhivotnovodstvo pozdnikh severian na slaviano-khazarskom pogranichie (Ch. 1)*. In: *Drevnosti 2014–2015*, Kharkov, 2015, 13, 212–227.

Krupskii and Polupan 1979 — Krupskii, N. K., and N. I. Polupan (eds.). *Atlas pochv Ukrainskoi SSR*. Kiev: Urozhai, 1979.

Kryganov 1989 — Kryganov, A. V. "Vooruzhenie i voisko naseleniia saltovo-maiackoi kul'tury (po materialam mogil'nikov s obriadom truposozhzheniia." In *Problemy arkheologii Podneprov'ia. Mezhvuzovskii sbornik nauchnykh trudo*. Edited by I. F. Kovaleva, 98–114. Dnepropetrovsk: Dnepropetrovskii gosudarstvennyi universitet.

Kuchera and Motsia 1999 — Kuchera, M. P. and O. P. Motsia. *Slov'iano-rus'ki gorodyshcha VIII–XIII st. mizh Sanom i Sivers'kym Dintsem*. Kiev: Institute of archaeology of NAS of Ukraine.

Kukharenko 1952 — Kukharenko, Iu. V. "Novopokrovskii mogil'nik i poseleniia." *Arkheolohiia* 6: 33–50.

Kunik and Rozen 1878 — Kunik, A. and V. Rozen. *Izvestiia al-Bekri i drugikh avtorov o Rusi i slavianakh*. St. Petersburg: Tipografiia Imperatorskoi akademii nauk.

Kuza 1996 — Kuza, A. V. *Drevnerusskie gorodishcha X–XIII vv.: svod arkheologicheskikh pamiatnikov*. Moscow: Khristianskoe izdatel'stvo.

Kvitkovskii 2008 — Kvitkovskii, V. I. "K probleme izucheniia i rekonstrukcii saltovskikh zhilishch lesostepnoi zony (na primere selishcha Korobovy Khutora Kharkovskoi obl.)." In *Slaviano-russkie drevnosti Dneprovskogo Levoberezh'ia. Materialy konferencii, posviashchennoi 75-letiiu so dnia rozhdeniia K. F. Sokola*. Edited by V. V. Enukov, 100–05. Kursk: Izdatel'stvo Kurskogo gosudarstvennogo universiteta.

Kvitkovskii 2009 — Kvitkovskii, V. I. "Istoriia arkheologicheskikh issledovanii mnogosloinogo selishcha Piatnitskoe-1 Pechenezhskogo raiona Kharkovskoi oblasti." *Slobozhans'ke kul'turne nadbannia*, no. 2: 85–91.

Kvitkovskii 2012 — Kvitkovskii, V. I. "Zhilishcha selishcha Korobovy Khutora." In *Stepi Evropy v epokhu srednevekov'ia. 9. Khazarskoe vrem'ia*. Edited by A. V. Evgelevskii, 99–118. Donetsk: Izdatel'stvo Doneckogo nacional'nogo universiteta.

Kvitkovskii, Pashkevich, and Gorbanenko 2011a — Kvitkovskii, V. I., G. A. Pashkevich and S. A. Gorbanenko. "K voprosu o zemledelii naseleniia saltovskogo selishcha Piatnickoe-I." In *Istoriia i arkheologiia vostochnoevropeiskoi derevni. Materialy I mezhdunarodnogo nauchnogo konferenciia "Problemy agrarnoi istorii i arkheologii Vostochnoi Evropy" (22–23 sentiabria 2011 g., s. Kokino)*, 99–15. Briansk: Brianskaia gosudarstvennaia sel'sko-khoziastvennaia akademii, Nauchno-issledovannye laboratoriia agrarnyi istorii.

Kvitkovskii, Pashkevich, and Gorbanenko 2011b — KvitkovskiI, V. I., G. O. Pashkevich, and S. A. Gorbanenko, S. A. "Materialy z zemlerobstva zhyteliv poselennia Piatnyts'ke-I." Arkheolohiia, no. 3: 122–128.

Lange 1975 — Lange, E. "The development of agriculture during the first millennium AD." *Geologiska Föreningens i Stockholm Förhandlingar* 97: 115–124.

Laptev 2007 — Laptev, A. A. "Srednevekovye okovki zastupov na Kharkovshchine." *Kharkovskii arkheologicheskii sbornik*, no. 2: 12–20.

Lebedeva 2008 — Lebedeva, E. Iu. "*Archeobotanicheskaia rekonstruktsiia drevnego zemledeliia (metodicheskiie kriterii)*." *Opus. Mezhdistsiplinarnye issledovaniia v arkheologii*, no. 6: 86–109.

Levashova 1956 — Levashova, V. P. "Selskoe khoziaistvo. Ocherki po istorii russkoi derevni." *Trudy Gosudarstvennogo istoricheskogo muzeia* 32: 19–103.

Liapushkin 1958a — Liapushkin, I. I. "Pamiatniki saltovo-maiatskoi kul'tury v basseine r. Dona." In *Trudy Volgo-Donskoi arkheologicheskoi ekspeditsii*. Edited by Mikhail I. Artamonov, 85–150. Moscow/Leningrad: Izdatel'stvo Akademii Nauk SSSR.

Liapushkin 1958b — Liapushkin, I. I. "Karnaukhovskoe poselenie." In *Trudy Volgo-Donskoi arkheologicheskoi ekspedicii*. Edited by Mikhail I. Artamonov, vol. 1, 263–314. Moscow/Leningrad: Izdatel'stvo Akademii Nauk SSSR.

Liapushkin 1958c — Liapushkin, I. I. *Gorodishche Novotroitskoe. O kul'ture vostochnykh slavian v period slozheniia Kievskogo gosudarstva*. Moscow/Leningrad: Izdatel'stvo Akademii Nauk SSSR.

Liapushkin 1959–1960 — Liapushkin, I. I. "K voprosu o pamiatnikakh Volyncevskogo tipa." *Sovetskaia Arkheologiia* 29–30: 58–83.

Likhachev 1982 — Likhachev, V. A. "Ralo iz pogrebeniia u s. Verkhniaia Maiovka." In *Drevnosti stepnogo Podneprovia (III–I tys. do n. e.). Sbornik nauchnykh trudov*. Edited by I. F. Kovaleva, 51–53. Dnepropetrovsk: Izdatel'stvo

Dnepropetrovskgo gosudarstvennogo Universiteta im. 300-letiia vossoedineniia Ukrainy s Rossiei.

Lobanova-Gulak and Tortika 2009 — Lobanova-Gulak, M., and A. A. Tortika. "Torgovlia Vostochnoi Evropy s musulmanskimi stranami i Dono-Donetskii rechnoi put' v seredine VIII–seredinie X vv.: istoriograficheskie tradicii i realnost'." In *Islam i suchasnyi svit. Roboty uchasnykiv drugogo ta tret'ogo vseukrainskykh konkursiv islamoznavchykh doslidzhen' molodykh uchenykh im. A. Krymskogo*, 167–196. Kiev: Ansar Foundation.

Logvinov and Shcherban' 1984 — Logvinov, K. T., and M. I. Shcherban' (eds.). *Priroda Ukrainskoi SSR. Klimat.* Kiev: Naukova dumka.

Ludwig 1982 — Ludwig, Dieter. "Struktur und Gesellschaft des Chazaren-Reiches im Licht der schriftlichen Quellen." Ph.D. dissertation, Westfälische Wilhelms-Universität. Münster.

Lysov 1968 — Lysov, V. N. *Proso.* Leningrad: Kolos.

Magomedov 1983 — Magomedov, M. G. *Obrazovanie Khazarskogo kaganata (po materialam arkheologicheskikh issledovanii i pis'mennym dannym).* Moscow: Nauka.

Magomedov 1987 — Magomedov, B. V. *Cherniakhovskaia kul'tura Severo-Zapadnogo Prichernomor'ia.* Kiev: Naukova dumka.

Makarenko 1906 — Makarenko, M. O. "Otchet ob arkheologicheskikh izsl'edovaniiakh v Kharkovskoi i Voronezhskoi guberniiakh v' 1905 g." *Izvestiia imperatorskoi arkheologicheskoi kommissii* 19: 117–56.

Makarenko 1992 — Makarenko, D. O. *Mykola Omelianovych Makarenko.* Kiev: Naukova dumka.

Maksimov and Petrashenko 1988 — Maksimov, E. V. and V. O. Petrashenko. *Slavianskie pamiatniki u s. Monastyrek na Srednem Dnepre.* Kiev: Naukova Dumka.

Mamonov 1952 — Mamonov, V. S. *Starinnyie orudiia dlia obrabotki pochvy iz s. Staroselie na Dnieprie.* In: *Sovietskaia etnografiia*, 1952, 4, pp. 67–90.

Mason 1995 — Mason, R. A. E. "The religious beliefs of the Khazars." *Ukrainian Quarterly* 60, no. 4: 383–415.

Matolcsi 1984 — Matolcsi, J. "Kosti zhivotnykh s gorodishcha, selishcha i mogil'nika (1978–1979 gg.)." In *Maiatskoe gorodishche. Trudy Sovetsko-Bolgaro-Vengerskoi ekspedicii.* Edited by Svetlana A. Pletneva, 237–70, 276–277. Moscow: Nauka.

Matviishyna 2010 — Matviishyna, Zh. M. (ed.). *Prostorovo-chasova koreliatsiia paleogeografichnykh umov chetvertynngho periodu na terytorii Ukrainy.* Kiev: Naukova dumka.

Matviishyna, Karmazynenko, Kvitkovs'kyi, and Zadverniuk 2012 — Matviishina, Zh. M., S. P. Karmazynenko, V. I. Kvitkovskyi, and G. P. Zadverniuk. "Rezultaty paleopedologichnykh doslidzhen' gruntiv i vidkladiv arkheologichnoi stoianky Saltivskoi kultury poblyzu s. Piatnyts'ke Kharkivskoi oblasti." *Naukovi zapysky Sumskogo derzhavnogo pedagogichnogo universytetu im. A. S. Makarenka. Geografichni nauky*, no. 3: 236–253.

Medvedenko 2006 — Medvedenko, N. A. *Istoriia i arkheologiia Khazarskogo kaganata v issledovanii M. I. Artamonova.* Voronezh: Istoki.

Miagkova 1998 — Miagkova, Iu. Ia. "Analiz osteologischeskogo materiala zi poselenii saltovo-maiackoi kul'tury." In *Problemy arkheologii iugo-vostochnoi Evropy. Tezisy dokladov VII Donskoi arkheologicheskoi konferencii*, 139–40. Rostov-na-Donu: n(o) p(ublisher).

Miagkova 2013 — Miagkova, Iu. Ia. "Osteologicheskii material poselenii i gorodishch Khazarskogo vremeni na Nizhnem Donu." In *Khazarskie drevnosti. Sbornik nauchnykh statei.* Edited by Vladimir V. Kliuchnikov and A. A. Kulakov, 262–297. Aksai: DonPechat.

Mikheev 1968 — Mikheev, V. K. "Osnovnye remeslennye proizvodstva bolgaro-alanskogo naseleniia Podonia v VIII–X v." Ph.D. dissertation, "Maksim Gorki" State University, Kharkov.

Mikheev 1985 — Mikheev, V. K. *Podon'e v sostave khazarskogo kaganata.* Kharkov: Vyshcha shkola.

Mikheev 1986a — Mikheev, V. K. "Sukhogomolshanskii mogil'nik." *Sovetskaia Arkheologiia*, no. 3: 158–173.

Mikheev 1986b — Mikheev, V. K. "Ekonomika i sotsialnye otnosheniia u naseleniia saltovo-maiatskoi kultury Podonia-Priazovia (seredina VIII–seredina X vv.)." Ph.D. dissertation. Institute of Archaeology, Kiev.

Miliutin 1909 — Miliutin, A. "Razkopki 1906 g. na Maiatskom' gorodishchy." *Izvestiia imperatorskoi arkheologicheskoi kommissii* 29: 153–63.

Milkov 1950 — Milkov, F. N. *Lesostep' Russkoi ravniny. Opyt landshaftnoi kharakteristiki.* Moscow: Izdatel'stvo Akademii nauk SSSR.

Minasian 1978a — Minasian, R. S. "Klassifikaciia ruchnogo zhernovogo postava (po materialam Vostochnoi Evropy I tysiacheletiia n.e.)." *Sovetskaia Arkheologiia*, no. 3: 101–12.

Minasian 1978b — Minasian, R. S. "Klassifikaciia serpov Vostochnoi Evropy zheleznogo veka i rannego srednevekov'ia." *Arkheologicheskii sbornik Gosudarstvennogo Ermitazha* 19: 74–85.

Miroliubov 1976 — Miroliubov, M. A. "Orudiia vtorichnoi obrabotki pochvy i uborki urozhaia iz Staroi Ladogi." *Arkheologicheskii sbornik Gosudarstvennogo Ermitazha* 17: 119–24.

Monin 1957 — Monin, A. S. *Geografiia pochv, s osnovami pochvovedeniia. Uchebnik dlia pedagogicheskikh institutov.* Moscow: Nauka.

Moskalenko 1965 — Moskalenko, A. N. *Gorodishche Titchikha. Iz istorii drevnerusskikh poselenii na Donu.* Voronezh: Izdatel'stvo Voronezhskogo universiteta.

Motsia 2003 — Motsia, O. P. (ed.). *Selo Kyivs'koi Rusi (za materialamy pivdennoru'skykh zemel').* Kiev: Shliakh.

Mykhaylyna 2014 — Mykhaylyna, L. P. "Novi materialy z Revnians'kogo gnizda poselen' raykovetskoi kul'tury." In *Vid venediv do Rusi. Zbirnyk naukovykh prats' na poshanu doktora istoryhcnykh nauk, profesora Denysa Nykodymovycha Kozaka z nagody iogo 70-littia*. Edited by Glib Iu. Ivakin, 357–364. Kiev: Maidan.

Mykhaylyna, Pashkevich, and Pyvovarov 2007 — Mykhaylyna, L. P., G. O. Pashkevich, and S. V. Pyvovarov. "Ril'nitstvo sloviano-rus'kogo naselennia mezhyrichchia verkhn'ogo Prutu ta seredn'ogo Dnistra." *Arkheolohiia*, no. 2: 57–66.

Noonan 1982 — Noonan, T. S. "Russia, the Near East, and the steppe in the early medieval period: an examination of the Sasanian and Byzantine finds from the Kama-Urals area." *Archivum Eurasiae Medii Aevi* 2: 269–302.

Noonan 1983 — Noonan, Th. S. "What does historical numismatics suggest about the history of Khazaria in the ninth century?" *Archivum Eurasiae Medii Aevi* 3: 265–81.

Noonan 1984 — Noonan, Th. S. "Why dirhems first reached Russia: the role of Arab-Khazar relations in the development of the earliest Islamic trade with Eastern Europe." *Archivum Eurasiae Medii Aevi* 4: 151–282.

Noonan 1985 — Noonan, Th. S. "Khazaria as an intermediary between Islam and Eastern Europe in the second half of the ninth century." *Archivum Eurasiae Medii Aevi* 5: 179–204.

Noonan 1992a — Noonan, Th. S. "Byzantium and the Khazars: a special relationship?" In *Byzantine Diplomacy. Papers from the Twenty-Fourth Spring Symposium of Byzantine Studies*. Edited by Jonathan Shepard and Simon Franklin, 109–132. Aldershot: Variorum.

Noonan 1992b — Noonan, Th. S. "Fluctuations in Islamic trade with Eastern Europe during the Viking Age." *Harvard Ukrainian Studies* 16: 237–59.

Noonan 1995–1997 — Noonan, Th. S. "The Khazar economy." *Archivum Eurasiae Medii Aevi* 9: 253–314.

Parfenov 1873 — Parfenov, P. D. "Pisma o sel'skom khoziaistve iugo-zapadnoi Rossii." *Russkii Vestnik* 106, no. 8: 635–669.

Pashkevich 1988 — Pashkevich, Galina A. "Paleobotanicheskie issledovaniia v oblasti slavianskoi arkheologii." In *Trudy V Mezhdunarodnogo Kongressa arkheologov-slavistov, Kiev 18–25 sentiabria 1985 g*. Edited by Volodymyr D. Baran, vol. 4, 169–174. Kiev: Naukova Dumka.

Pashkevich 1991a — Pashkevich, G. A. *Paleoetnobotanicheskiie nakhodki na territorii Ukrainy (neolit-bronza.): Katalog*. Kiev: Institut Arkheologii AN Ukrainy.

Pashkevich 1991b — Pashkevich, G. A. *Paleoetnobotanicheskie nakhodki na territorii Ukrainy: pamiatniki I tys. do n. e.–II tys. n. e. Katalog I*. Kiev: Institut Arkheologii AN Ukrainy.

Pashkevich 1991c — Pashkevich, G. A. *Paleoetnobotanicheskie nakhodki na territorii Ukrainy: pamiatniki I tys. do n. e.–II tys. n. e. Katalog II*. Kiev: Institut Arkheologii AN Ukrainy.

Pashkevich 1991d — Pashkevich, G. A. *Paleoetnobotanicheskie nakhodki na territorii Ukrainy: drevniaia Rus'. Katalog*. Kiev: Institut Arkheologii AN Ukrainy.

Pashkevich 1992 — Pashkevich, G. A. "Kulturnyie rasteniia Ukrainy ot neolita do srednevekov'ia (po paleoetnobotanicheskim materialam)." Ph.D. dissertation, Institute of botany of Academy of Sciences of Ukraine. Kiev.

Pashkevich 1993 — Pashkevich, G. O. "Zminy v systemi zernovogo gospodarstva Davn'oi Rusi (chas, prychyny ta naslidky)." In *Starozhytnosti Pivdennoi Rusi. Materialy III istoryko-arkheologichnogo seminaru "Chernihiv i iogo okruga v IX–XIII st.", Chernihiv, 15–18 travnia 1990 r.* Edited by P. P. Tolochko, 87–91. Chernihiv: Siverians'ka dumka.

Pashkevich 2002 — Pashkevich, G. O. "Paleobotanichni doslidzhennia v Ukraini." In *Novi tekhnologii v arkheologii. Zbirnyk naukovykh prats'*. Edited by Denis N. Kozak, 282–293. Kiev/L'viv: Instytut arkheologii NAN Ukrainy/Instytut Arkheologii LNU im. I. Franka.

Pashkevich 2005 — Pashkevich, G. O. "Arkheolohiia ta paleoetnobotanika." *Arkheolohiia*, no. 2: 80–88.

Pashkevich and Gorbanenko 2002–2003 — Pashkevich, G. O., and S. A. Gorbanenko. "Vidbytky zernivok kul'turnykh roslyn na keramitsi Opishnianskogo gorodishcha." *Arkheologichnyi litopys Livoberezhnoi Ukrainy*, no. 1: 161–163.

Pashkevich and Gorbanenko 2004 — Pashkevich, G. O. and S. A. Gorbanenko. "K voprosu o zemledelii plemen saltovskoi kul'tury (po materialam poseleniia Rogalik)." *Drevnosti*: 225–236.

Pashkevich and Videyko 2006 — Pashkevich, G. O., and M. Iu. Videyko. *Ril'nitstvo plemen tripil's'koi kul'tury*. Kiev: Instytut arkheologii NAN Ukrainy.

Pidoplichko 1937 — Pidoplichko, I. G. "Do pytannia pro sviiski tvaryny tripil's'kikh poselen Khalepia, Andriivka, Usatove." *Naukovi zapysky Instytutu istorii material'noi kul'tury*, no. 2: 111–120.

Pletneva 1967 — Pletneva, S. A. *Ot kochevyi k gorodam: saltovo-maiatskaia kul'tura*. Moscow: Nauka.

Pletneva 1984 — Pletneva, S. A. "Risunki na stenakh Maiackogo gorodishcha." In *Maiatskoe gorodishche. Trudy Sovetsko-Bolgaro-Vengerskoi ekspeditsii*. Edited by Svetlana A. Pletneva, 57–94, 273–274. Moscow: Nauka.

Pletneva 1986 — Pletneva, S. A. *Khazary*. Moscow: Nauka.

Pletneva 1989 — Pletneva, S. A. *Na slaviano-khazarskom pogranich'e. Dmitrievskii arkheologicheskii kompleks*. Moscow: Nauka.

Pletneva 1990 — Pletneva, S. A. "Khazarskie problemy v arkheologii." *Sovetskaia Arkheologiia*, no. 2: 77–91.

Pletneva 1994 — Pletneva, S. A. "Pravoberezhnye Tsimlianskoe gorodishche. Raskopki 1958–1959 gg." *Materialy po arkheologii, istorii i etnografii Tavrii* 4: 271–396.

Pletneva 1996 — Pletneva, S. A. *Sarkel i "shelkovyi" put'*. Voronezh: Voronezhskii Gosudarstvennyi universitet.

Pletneva 1999 — Pletneva, S. A. *Ocherki khazarskoi arkheologii*. Moscow/Jerusalem: Mosty kul'tury/Gesharim.

Pokrovskii 1905 — Pokrovskii, A. M. "Verkhnesaltovskii mogil'nik." In *Trudy XII Arkheologicheskago săezda v Kharkove 1902 g.*, vol. 1, 465–91. Moscow: Tovarishchestvo tipografii A. I. Mamontov.

Ponomarev 1955 — Ponomarev, N. A. *Istoriia tekhniki mukomol'nogo i krupianogo proizvodstva*, vol. 1. Moscow: Zagotizdat.

Ponomarev 2012 — Ponomarev, L. Iu. "Khoziaistvennaia deiatel'nost' naseleniia saltovskoi kul'tury Kerchenskogo poluostrova (kratkii obzor arkheologicheskikh istochnikov)." *Saltovo-maiats'ka arkheologichna kul'tura: problemy ta doslidzhennia*, no. 2: 67–79.

Popov, Marynych, and Lan'ko 1968 — Popov, V. P., A. M. Marynych, and A. I. Lan'ko (eds.). *Fiziko-geograficheskoe raionirovanie Ukrainskoi SSSR*. Kiev: Izdatel'stvo Kievskogo universiteta, 1968.

Potapov 1990 — Potapov, O. V. "Novye materialy k istorii skotovodstva i okhoty v drevnem Chernigove." In *Problemy arkheologii Iuzhnoi Rusi. Materialy istorikoi-arkheologicheskoi seminara "Chernigov i ego okruga v IX–XIII vv.", Chernigov, 26–28 sent. 1988 g.* Edited by P. P. Tolochko, O. P. Motsia, R. S. Orlov, A. L. Kazakov and V. P. Kovalenko, 37–41. Kiev: Naukova dumka.

Pritsak 1981 — Pritsak, O. *The Origin of Rus'*. Cambridge, Mass.: Harvard University Press.

Pritsak 1995 — Pritsak, O. "The system of government under Volodimer the Great and his foreign policy." *Harvard Ukrainian Studies* 19: 573–93.

Pugolovok 2014 — Pugolovok, Iu. O. Review of *Sil's'ke gospodarstvo na slov'iano-khozarskomu porubizhzhi*, by S. A. Gorbanenko and V. V. Koloda (Kiev, 2013). *Arkheolohiia*, no. 3: 121–24.

Pyvovarov and Il'kiv 2014 — Pyvovarov, S. V. and M. V. Il'kiv. "Ridkivs'ke poselennia raykovets'koi kul'tury u Verkhn'omu Poprutti." In *Vid venediv do Rusi. Zbirnyk naukovykh prats' na poshanu doktora istoryhcnykh nauk, profesora Denysa Nykodymovycha Kozaka z nagodu iogo 70-littia*. Edited by Glib Iu. Ivakin, 345–356. Kiev: Maidan.

Rybakov 1962 — Rybakov, B. A. "Kalendar IV v. iz zemli polian." *Sovetskaia arkheologiia*, no. 4: 66–89.

Serbina 1950 — Serbina, K. N. (ed.), *Kniga Bol'shomu chertezhu*. Moscow: Izdatel'stvo Akademii nauk SSSR.

Shramko 1964 — Shramko, B. A. "Drevnii dereviannyi plug iz Sergeevskogo torfianyka." *Sovetskaia arkheologiia*, no. 4: 96–99.

Shramko 1972 — Shramko, B. A. "Pro chas poiavy ornogo zemlerobstva na pivdennomu skhodi Evropy." *Arkheolohiia*, no. 7: 25–35.

Shramko 1983 — Shramko, B. A. "Pogrebenie VIII–X vv. u s. Piatnickoe v Kharkovskoi obl." In *Drevnerusskoe gosudarstvo i slaviane. Materialy simpoziuma posviashchaeschie 1500-letiiu Kieva*. Edited by Leonid D. Pobol', 48–50. Minsk: Nauka i tekhnika.

Shramko 1984 — Shramko, B. A. "Model bychei upriazhki skifskogo vremeni." In *Drevnosti Evrazii v skifo-sarmatskoe vremia*. Edited by A. I. Meliukova, M. G. Moshkova, and V. G. Petrenko, 231–255. Moscow: Nauka.

Shramko 1996 — Shramko, B. A. "Kompleks glinianykh skulptur Bel'skogo gorodishcha." In *Bil'ske gorodyshche v konteksti vyvchennia pam'iatok rann'ogo zaliznogo viku Evropy*. Edited by P. K. Bondaresvs'kyi and O. Suprunenko, 67–87. Poltava: TSODPA.

Shvets' 1946 — Shvets', G. I. *Rozpodil stoku richok URSR po sezonakh i misiatsakh*. Kiev: Vydavnytstvo Akademii Nauk URSR.

Shvets', Drozd, and Levchenko 1957 — Shvets', G. I., N. I. Drozd, and S. P. Levchenko. *Katalog richok Ukrainy*. Kiev: Vydavnytstvo Akademii Nauk URSR.

Skrzhinskaia 1971 — Skrzhinskaia, E. Ch. (ed.), *Barbaro i Kontarini o Rossii*. Leningrad: Nauka.

Sleptsov and Klimenko 2005 — Sleptsov, A. M., and V. V. Klimenko. "Obobshchenie paleoklimaticheskikh dannykh i rekonstruktsiia klimata Vostochnoi Evropy za posledniie 2000 let." *Istoriia i sovremennost'*, no. 2: 118–137.

Slobodin 1952 — Slobodin, V. M. "K voprosu o razvitii i smene sistem zemledeliia (ot vremen ego vozniknoveniia na territorii SSSR do I tys. n. e.)." In *Materialy po istorii zemledeliia SSSR*, vol. 1, 9–65. Moscow: Akademia nauk SSSR, Institut istorii.

Smirnov and Sosnikhina 1984 — Smirnov, V. R., and S. P. Sosnikhin. *Genetika rzhi*. Leningrad: Izdatel'stvo Leningradskogo gosudarstvennogo universiteta.

Sorokin 1959 — Sorokin, S. S. "Zheleznye izdeliia Sarkela-Beloi Vezhi." In *Trudy Volgo-Donskoi arkheologicheskoi ekspeditsii*. Edited by Mikhail I. Artamonov, vol. 2, 135–99. Moscow/Leningrad: Nauka.

Spitsyn 1909 — Spitsyn, A. A. "Istoriko-arkheologicheskie izyskaniia. Iskonnye obitateli Dona i Dontsa." *Zhurnal Ministerstva narodnogo prosveshcheniia* 1: 67–98.

Streuver 1968 — Streuver, S. 1968. "Flotation techniques for the recovery of small-scale archaeological remains." *American Antiquity* 33: 353–362.

Suess 1978 — Suess, H. E. "Natural radiocarbon, solar activity, and climate." *Geological Survey, Open-file Report* 71.

Sukachev 1940 — Sukachev, V. N. *Opredelitel' drevesnykh porod*. Moscow: Goslestekhizdat.

Svistun 2012 — Svistun, G. E. "Novyi krematsionnyi mogil'nik na territorii Chuguevo-Babchanskogo lesnichestva (predvaritel'naia informatsiia)." *Saltovo-maiats'ka arkheologichna kul'tura: problemy ta doslidzhennia*, no. 2: 79–85, 133–141.

Svistun and Chendev 2002–2003 — Svistun, G. E., and Iu. G. Chendev. "Vostochnyi uchastok oborony Mokhnachanskogo gorodishcha i ego prirodnoe okruzhenie v drevnosti." *Arkheolohichnyi litopys Livoberezhnoi Ukrainy*: 130–135.

Svistun and Gorbanenko 2011 — Svistun, G. E. and S. A. Gorbanenko. "Spozhyvchi aspekty sotsial'no-ekonomichnykh vidnosyn na prykladi Chuguivs'kogo gorodyshcha saltivs'kogo chasu." *Saltovo-maiats'ka arkheologichna kul'tura: problemy ta doslidzhennia*, no. 1: 75–86, 136–143.

Sycheva and Chichagova 1999 — Sycheva, S. A., and O. A. Chichagova. "Radiouglerodnaia khronostratigrafiia golotsenovykh otlozhenii Srednerusskoi vozvyshennosti." *Biulleten Komissii po izucheniiu chetvertichnogo perioda Rossiiskoi akademii nauk* 63: 104–113.

Symonovych 1952 — Symonovych, E. O. "Mlynove sporudzhennia pershykh stolit nashoi ery." *Arkheolohiia*, no. 6: 97–107.

Tikhonenko 1995 — Tikhonenko, D. G. *Gruntoznavstvo chastkove*. Kharkiv: without publisher.

Timchenko 1972 — Timchenko, N. G. *K istorii okhoty i zhivotnovodstva v Kievskoi Rusi (Sredniee Podneprov'e)*. Kiev: Naukova dumka.

Tomashevskyi 2003 — Tomashevskyi, A. P. "Narys istorii ta ekologii zaselennia Skhidnogo Podillia sloviano-ruskoi doby." *Arkheologichni studii* 2: 132–160.

Tomashevskyi 2008 — Tomashevskyi, A. P. "Informatsiina osnova doslidzhennia istorii ta ekologii zaselennia Skhidnoho Podillia sloviano-ruskoi doby." *Arkheologichni studii* 3: 169–179.

Tomashevskyi 2010 — Tomashevskyi, A. P. "Prostorovi ta paleoekologichni doslidzhennia Pivdennoru'skykh zemel': rezul'taty, novi pidkhody, perspektyvy." *Arkheologiia i davnia istoriia Ukrainy*, no. 1: 174–206.

Tortika 2006a — Tortika, A. A. "Dono-Donetskii rechnyi put', lesostepnoe Pridonech'e i vizantiiskaia Tavrika v VIII–X vv.: kharakter i napravleniia torgovykh sviazei." In *Bospor Kimmeriiskii i varvarskii mir v period antichnosti i srednevekov'ia: Oikos*. Edited by V. N. Zin'ko, 282–86. Kerch': Tsentr arkheologicheskikh issledovanii.

Tortika 2006b — Tortika, A. A. *Severo-zapadnaia Khazariia v kontekste istorii Vostochnoi Evropy*. Kharkiv: Khar'kovskaia Gosudarstvennaia Akademiia Kul'tury.

Tortika 2006c — Tortika, A. A. "Torgovlia s musulmanskimi stranami v Vostochnoi Evrope v VIII–X vv." In *Materialy trinadtsatoi ezhegodnoi mezhdunarodnoi mezhdistsiplinarnoi konferentsii po iudaike*. Edited by K. Iu. Burmistrov, 319–40. Moscow: Sefer.

Tret'iakov 1932 — Tret'iakov, P. N. *Podsechnoe zemledelie v Vostochnoi Evrope*. Leningrad: Pechatnyi dvor.

Tret'iakov 1946 — Tret'iakov, Petr N. "Vostochnoslavianskie plemena v svete arkheologicheskikh issledovanii poslednikh let." *Kratkie soobshcheniia o dokladakh i polevykh issledovaniiakh Instituta istorii material'noi kul'tury AN SSSR* 1: 39–46.

Tret'iakov 1947 — Tret'iakov, Petr N. "Starodavni slov'ians'ki gorodishcha u verkhnii techii Vorskla." *Arkheolohiia*, no. 1: 123–140.

Tret'iakov 1951 — Tret'iakov, Petr N. "Sel'skoe khoziaistvo i promysly." In *Istoriia kul'tury drevnei Rusi*. Edited by Nikolai N. Voronin, Mikhail K. Karger and Mariia A. Tikhanova, vol. 1, 47–77. Moscow/Leningrad: Izdatel'stvo Akademii Nauk SSSR.

Tret'iakov 1966 — Tret'iakov, Petr N. *Finno-ugry, balty i slaviane na Dnepre i Volge*. Moscow/Leningrad: Nauka.

Tret'iakov 1969 — Tret'iakov, Petr N. "Nekotorye itogi izucheniia vostochnoslavianskikh drevnostei." *Kratkie soobshcheniia o dokladakh i polevykh issledovaniiakh Instituta istorii material'noi kul'tury AN SSSR* 118: 20–31.

Tsalkin 1956 — Tsalkin, V. I. *Materialy dlia istorii zhivotnovodstva i okhoty v drevnei Rusi*. Moscow Izdatel'stvo Akademii Nauk SSSR.

Tsalkin 1969 — Tsalkin, V. I. "Fauna iz raskopok borshevkikh i romenskikh go rodishch." *Sovetskaia Arkheologiia*, no. 4: 91–101.

Vashchenko 2006 — Vashchenko, E. D. *"Khazarskaia problema" v otechestvennoi istoriografii XVIII–XX vv.* ("The Khazar problem" in modern historiography, from the eighteenth to the twentieth century). St. Petersburg: Izadatel'stvo Sankt-Peterburgskogo universiteta.

Vasiulina 1970 — Vasiulina, O. D. (ed.). *Buriany Ukrainy (Vyznachnyk-dovidnyk)*. Kiev: Naukova dumka.

Vavilov 1986 — Vavilov, P. P. (ed.), *Rastenievodstvo*. Moscow: Agropromizdat.

Veklych 1987 — Veklich, M. F. *Problemy paleoklimatologii*. Kiev: Naukova dumka.

Veklych and Gerasimenko 1993 — Veklych, M. F., and N. P. Gerasimenko. "Etapnist seredn'o-pizn'ogolotsenovogo grunto- i porodoutvorennia na zaplavi Nyzhn'oi Desny." *Visnyk Kyivskogo universytetu: Seriia Geografiia* 40: 85–92.

Veretiushkina 2008 — Veretiushkina, M. V. "Stanovlenie i razvitie zemledeliia slavian na territorii mezhdurech'ia Seima i Psla (po nakhodkam nakonechnikov pashennykh orudii)." In *Slaviano-russkie drevnosti Dneprovskogo Levoberezh'ia. Materialy konferencii, posviashchennoi 75-letiiu so dnia rozhdeniia K. F. Sokola*. Edited by V. V. Enukov, 29–34. Kursk: Izdatel'stvo Kurskogo gosudarstvennogo universiteta.

Veretiushkina 2011 — Veretiushkina, M. V. "Klad sel'skokhoziaistvennykh orudii iz derevni Sugrovo." *Vestnik Voronezhskogo gosudarstvennogo universiteta. Seriia Istoriia, politologiia, sotsiologiia*, no. 2: 84–88.

Vernander 1963 — Vernander, N. B. "Proiskhozhdeniie i svoistva serykh lesnykh pochv zapadnoi chasti USSR." In *Issledovaniia v oblasti genezisa pochv*. Edited by E. N. Ivanova and N. A. Nogina, 164–183. Moscow: Izdatel'stvo Akademii Nauk SSSR.

Vernander and Tiutiunnik 1986 — Vernander, P. B., and D. A. Tiutunnik (eds.). *Priroda Ukrainskoi SSR. Pochvy*. Kiev: Naukova dumka.

Veselovskii, Lysenko, and Man'ko 1988 — Veselovskii, I. V., A. K. Lysenko, and Iu. P. Man'ko. *Atlas-vyznachnyk burianiv*. Kiev: Urozhai.

Viazov 2001 — Viazov, A. "Sistema rasseleniia plemen imen'kovskoi kul'tury v podgornoi chasti Samarskoi luki." In *Samarskii krai v istorii Rossii. Materialy iubilenoi nauchnoi konferencii, 6–7 fevralia 2001 g.* Edited by Petr S. Kabytov, 55–69. Samara: Fain dizain.

Vikhrov 1959 — Vikhrov, V. E. *Diagnosticheskie priznaki drevesiny glavneishikh lesokhoziaistvennykh i lesopromyshlennykh porod SSSR*. Moscow: Akademiia Nauk SSSR.

Vinnikov 1984 — Vinnikov, A. Z. "Zhilye i khoziaistvennye postroiki Maiatskogo selishcha (rezul'taty raskopok 1975, 1977, 1978 gg.)." In *Maiatskoe gorodishche. Trudy Sovetsko-Bolgaro-Vengerskoi ekspeditsii*. Edited by Svetlana A. Pletneva, 95–135, 274. Moscow: Nauka.

Vinnikov 1995 — Vinnikov, A. Z. *Slaviane lesostepnogo Dona v rannem srednevekov'e (VIII–nachalo XI veka)*. Voronezh: Izdatel'stvo Voronezhskogo universiteta.

Vinnikov 2012 — Vinnikov, A. Z. "Izdeliia iz zheleza s Zhivotinnogo gorodishcha borshevskoi kul'tury (VIII–nach. XI vv.)." *Divnogorskii sbornik*, no. 3: 123–141.

Vinnikov 2014 — Vinnikov, A. Z. *Iugo-vostochnaia okraina slavianskogo mira v VIII–nach. XIII vv. (Zhivotinnoe gorodishche na r. Voronezh)*. Voronezh: Kvarta.

Vinnikov and Gorbanenko 2013 — Vinnikov, A. Z. and Sergei A. Gorbanenko. "Sel'skoe khoziaistvo zhitelei Zhivotinnogo gorodishcha." *Stratum+*, no. 5: 327–358.

Vinnikov and Pletneva 1998 — Vinnikov, A. Z. and S. A. Pletneva. *Na severnyk rubezhakh Khazarskogo kaganata. Maiatskoe poselenie*. Voronezh: Izdatel'stvo Voronezhskogo gosudarstvennogo universiteta.

Vinnikov and Stepovoi 2012 — Vinnikov, A. Z. and A. V. Stepovoi. *Drevnie metallurgi Pooskol'ia (Iutanovskii metallurgicheskii kompleks saltovo-maiackoi kul'tury)*. Voronezh: Izdatel'stvo Voronezhskogo gosudarstvennogo universiteta.

Voevodskii 1949 — Voevodskii, M. V. "Gorodishcha verkhnei Desny." *Kratkie soobscheniia Instituta istorii material'noi kul'tury* 24: 67–77.

Zelenin 1907 — Zelenin, D. K. *Russkaia sokha, eia istoriia i vidy. Ocherk iż istorii russkoi zemledel'cheskoi kul'tury*. Viatka: Izdatel'stvo Viatskago gubernskago stat. komiteta.

Zelenin 1991 — Zelenin, D. K. *Vostochnoslavianskaia etnografiia*. Moscow: Nauka.

Zhuravlev 1991 — Zhuravlev, O. P. "Zhivotnovodstvo i okhota u plemen epokhi bronzy na territorii Severnogo Prichernomoria i Priazovia." In *Drevneishie obshchnosti zemledel'tsev i skotovodov Severnogo Prichernomor'ia v tys. do n. e. v v. n. e.* Edited by E. V. Iarovoi and T. A. Shcherbakova, 137–138. Tiraspol: Tsentr arkheologicheskikh issledovanii.

Zhuravlev 1998 — Zhuravlev, O. P. "Zhivotnovodstvo u slavianskogo naseleniia vostochnoevropeiskoi lesostepi vo vtoroi polovine I tysiacheletiia nashei ery." In *Voprosy istorii slavian. Arkheologiia i etnografiia. Sbornik nauchnykh trudov*. Edited by A. Z. Vinnikov, V. I. Dynin, N. P. Mananchikova and M. V. Cybin, 34–43. Voronezh: Izdatel'stvo Voronezhskoi universiteta.

Zhuravlev 2001 — Zhuravlev, O. P. *Osteologicheskie materialy iz pamiatnikov epokhi bronzy lesostepnoi zony Dnepro-Donskogo mezhdurechia*. Kiev: Institut arkheologii NAN Ukrainy.

Zuckerman 1995 — Zuckerman, C. "On the date of the Khazars' conversion to Judaism and the chronology of the kings of the Rus, Oleg and Igor. A study of the anonymous Khazar letter from the Genizah of Cairo." *Revue des études byzantines* 53: 237–70.

Index

www.ingramcontent.com/pod-product-compliance
Lightning Source LLC
Chambersburg PA
CBHW061221240926
51846CB00006B/1305